WE GOT STEAM HEAT!

A Homeowner's Guide to Peaceful Coexistence

For additional copies, contact
HeatingHelp.com
63 North Oakdale Avenue,
Bethpage, NY 11714
Telephone: 1-800-853-8882
Fax: 1-888-486-9637
www.HeatingHelp.com

Manufactured in the U.S.A.

First Printing, October 2003
Second Printing, January 2006
Third Printing, December 2007

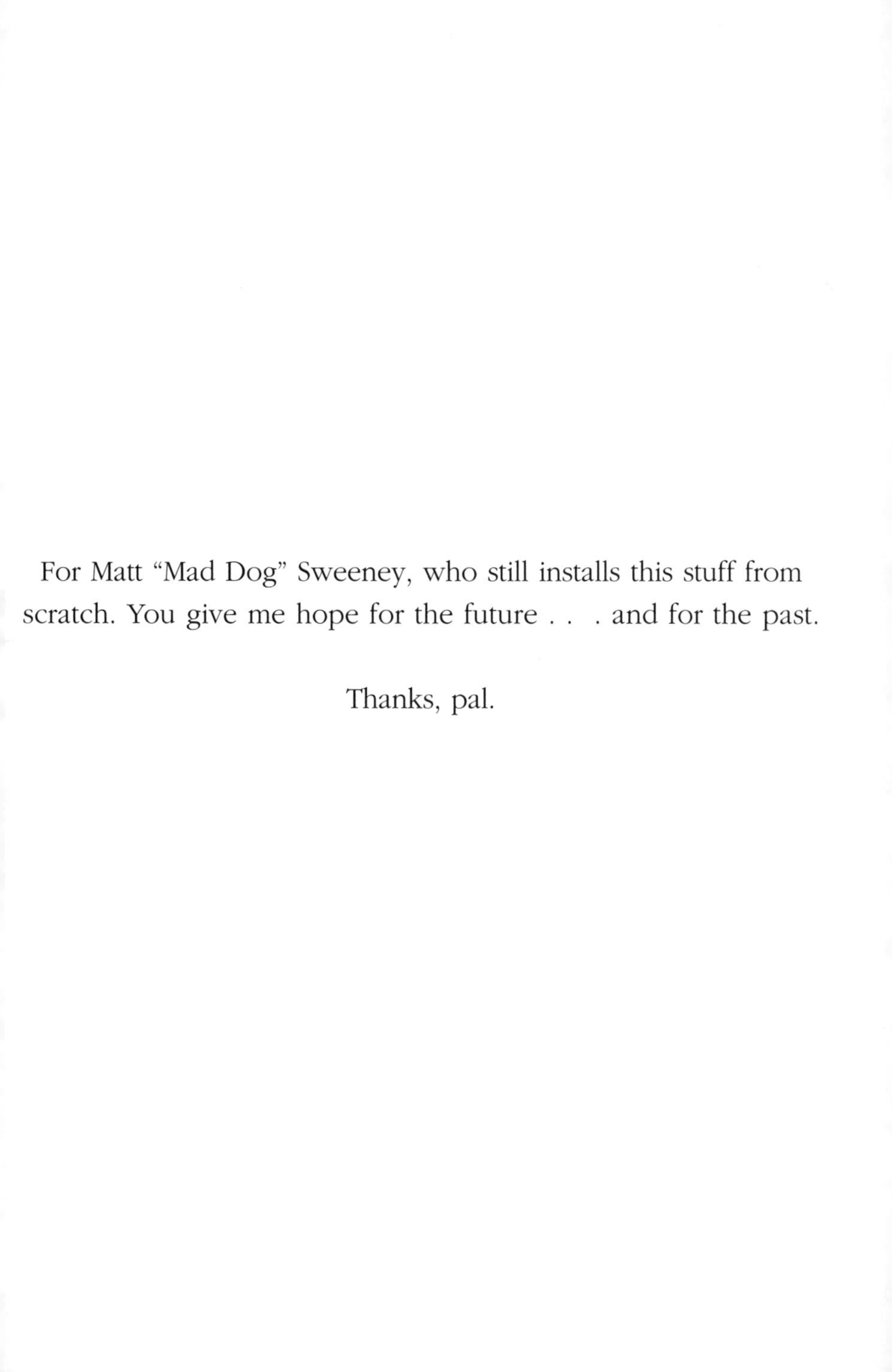

For Matt "Mad Dog" Sweeney, who still installs this stuff from scratch. You give me hope for the future . . . and for the past.

Thanks, pal.

So you got steam heat?

Good for you! You own a piece of American history there, along with radiators that weigh more than most midsize cars and pipes that let you know they're in the house. Ahh, history!

When working well, your radiators probably glow gently, and warm you wonderfully, and give you a place to put those wet woolen hats and mittens. And the cat and the dog certainly love those radiators. What better place to doze away the day than in front of a few hundred pounds of decorative cast iron?

When not working well, however, your system probably growls and hammers and knocks and complains like a nasty old man. I'll bet it even spits dirty water on your walls and curtains. And it probably uses more fuel than an Army convoy and causes your neighbors to ask you why you don't just do something about it because the hammering in your house is keeping them awake in *their* house. And it's all so grand and so confusing, and all at the same time.

Yes, you got steam heat, and you figure you're stuck with

it. And you probably haven't done anything about it (at least to this point) because you figure that most of the people who understand steam heat nowadays also happen to be dead. And you don't know what to do, or where to turn.

But fear not because you and I are in this together now, and I *love* this stuff. And I'm going to tell you things that will make your imagination soar. You'll see. And you're going to realize that steam heating really isn't all that complicated.

It's not. Really!

I'm going to explain it all to you in plain English. I'm going to show you that there's no reason to tear your house apart, no reason to rip out all those pipes and wonderful old radiators. There really isn't. I'm going to show you how to tame the beast and make it as gorgeous as it once was, and who knows, all of this might even turn into a new hobby for you. Steam heating can be quite intriguing. You'll see.

So we begin. Together.

Right now I figure . . .

✓ You want to understand how your steam heating system works.

✓ You want to know why it's making those strange noises, why it goes bump in the night (and sometimes all day long too!).

✓ You want to save money on fuel (and who doesn't?).

✓ You want to understand what all those strange metal objects hanging from your basement ceiling and poking up from the pipes are. And you wonder what might happen if you touch them, or (Yikes!) have them removed.

✓ And you may be wondering if you can move some of the pipes in your basement so you don't have to wear a hard hat when you go downstairs to do the laundry.

✓ You might want to know how to find a knowledgeable steam heating contractor, and how to avoid the knuckleheads who masquerade as knowledgeable contractors.

✓ And if you're a daring Do-It-Yourselfer, you probably want to know how to fix all of this stuff, and I'll tell you a lot about that, but I'll also tell you when I think you should keep your hands in your pockets and leave it to the pros.

✓ And I'm sure you have lots of other questions but don't worry; we'll get to them all before we're done. Promise.

So. Is it normal for steam heating systems to knock and bang and spit and keep you up all night and make you nuts during the day?

No, it isn't!

And can you make all of these problems go away?

Yes, you can!

And will it be expensive? Well, it might be. It depends. And I'm not trying to be evasive when I say that. It *does* depend. It depends on a lot of things. The best way to find out what you're up against is to understand the system that's in your own home. And that's what I'm here to do. To explain it all in plain English. You ready?

Great! Let's get to it.

First, some basic stuff

Beware of the knuckleheads!

You're having a problem with your steam system. It's not heating all the rooms in your home and it's knocking and ticking and keeping you up at night, and you don't know what to do so you call a local heating contractor. He shows up and slogs down your basement stairs. He finds your old boiler and stares at it for a while. He gawks up and around at the many pipes that run here and there, and at the dark, cast iron objects that hang from those pipes. He mutters to himself. You notice that he does that a lot – the muttering.

Finally, he looks at the pressure gauge on your boiler. He taps it with a knuckle. "What pressure you running?" he asks, and you have no idea what he's talking about. You shrug. He smirks. "You need a good head of steam for a house this size," he declares. You're thinking your house really isn't that big. You shrug again, waiting for the wisdom that you're hoping will come. "We gotta get some more pressure here," he says. "That's what she needs. More pressure."

And then he takes out this special screwdriver that turns in only one direction, that being clockwise, and he applies his tool to a device on your boiler that we call the **pressuretrol** (that's short for "pressure control," and I'll tell you all about it in a little while). He cranks the screw on the pressuretrol all the way down and he nods confidently. "Gotta have enough pressure to push the steam up to those bedrooms on the second floor," he says, and then he brings your boiler to a pressure that would be appropriate for the Empire State Building. He does this because he is a knucklehead.

But you're not aware of that yet.

And you know what? I'll bet what the knucklehead is doing even makes sense to you at this point in your life because you're just a few pages into this book, and what the heck do you know about steam heat? That's why you bought this book, right?

I used to think the same as you're probably thinking right now. You need a good head of steam to get all the way up to those top-floor radiators! I grew up reading Mark Twain and dreaming of steamboats and I watched all those cowboy movies with all those steam locomotives, and that's power, man, all that steam screaming from whistles and that's what we need to get up to the second-floor radiators. Power!

You see, I, too, was once a knucklehead.

And it all made so much sense to me back then. Raise the pressure and things should start happening. Simple! The steam won't go upstairs? "Oh, *I'll* get it upstairs," I'd say, and follow my bold statement with a manly, knowing nod.

I was once a knucklehead.

So that's what the knucklehead in your basement does. He cranks up the pressure on your boiler and he gives you a bill, which you pay, and he leaves.

And you will now have even higher fuel bills than you had before you met him, and the problems you had before he arrived will be a bit worse than they were in the pre-knucklehead days.

And you paid him. And maybe you're feeling like a knucklehead too at that point. You paid him.

Pressure kills

If you take away one thing from this entire book, I'm hoping that it's the firm conviction that your steam heating system is *not* a steam engine. Nor is it a steam locomotive, and neither is it the Delta Queen chugging up the Missouri River.

The steam heating system in your home doesn't need a good head of steam. All it needs is a gentle pressure to nudge the air from the pipes so that the steam can do its work. In most cases, all you need is a few ounces of pressure. Anything more is wasteful, and harmful.

There are many good steam heating contractors out there, and I'm going to be encouraging you throughout this book to seek out the good guys for much of this work. There are certain things that I think you, as a homeowner, can handle, but other things are best left to the knowledgeable pros. You will recognize these people by the way they view pressure in a steam heating system. Run from the guy who tells you that you need a good head of steam to heat a house. You don't.

And there's a real good reason why the Dead Men (those are the pros who installed this stuff back in the day) designed house-heating steam systems to run at low pressure. Here, I need to share this with you early on. It's a touching, and very telling, story, one whose theme repeated itself all too often in American history. This one, the story of Anthony Evans, engineer, and Alberta Williard, a schoolgirl of 12, appeared in the *Chicago Tribune* on January 14, 1886. The dateline was Fort Wayne, IN.

Listen.

Saint Mary's Catholic Church, on Lafayette Street, was entirely destroyed today by the explosion of the steam boiler of the steam-heating apparatus. The walls and tower are still standing, but by the order of the Mayor and Council will be taken down at once to prevent further fatalities. Anthony Evans, the engineer, was instantly killed. He was in the boiler room at the time, engaged in getting the church warmed, to have it ready for about 200 schoolchildren, who were to attend services later in the afternoon. Evan's remains were buried in the debris, and the firemen were over an hour digging them out. Alberta Williard, a schoolgirl of 12, had just left her father's house, nearby, on her way to school, and had reached the front of the church when the explosion took place. One of the large entrance doors was hurled outward with great force, striking her on the head, killing her instantly.

The church was built around twenty years since, at a cost of $30,000, and improvements to the amount of $30,000 have been added, making a total of $60,000, on which there is an insurance of $24,000. There is some doubt expressed about it being collectable. The boiler was entirely new. The entire heating apparatus was put in last September. Pipes were distributed all over the church, and ran under nearly every pew, which caused the explosion to be a general one. Parts of the boiler were thrown 200 feet in the air, and the explosion was felt all over the city.

A schoolgirl of 12. Killed by a church door. The engineer was in the boiler room, preparing the building for the children. He's killed instantly. Can you imagine the boiler pressure required to cause this much sudden devastation? An explosion felt all over the city? And events such as the disaster at Saint Mary's were so common in those days that they were hardly news anymore. This *Chicago Tribune* story was buried in the back pages – just a minor event in 19th Century America.

By the turn of the 20th Century, though, people had had enough and the heating industry began to establish standards that would help prevent such horrible explosions. The most important of these standards called for a maximum pressure of two pounds per square inch (**psi**) for heating boilers in homes and public buildings. These standards set up the use of properly sized pipes that would allow a heating system to run well at a very low (and far safer) pressure. And that's why the steam pipes in your home are so large, and why your boiler never has to produce a pressure greater than two-psi to work, no matter how large your house is. All you need is two-psi, and almost always, you can get by with less. It's true.

And if the knucklehead tries to set your boiler at a pressure greater than two pounds per square inch, just show him this book, and tell him that he's missing something.

What he's probably missing

When water boils and turns to steam it expands about 1,700 times in volume. Think of it! You would need 1,700 empty quart bottles to collect the steam that comes off of a single quart of boiling water. Isn't that amazing? It's this sudden expansion, this instant propulsion, that causes the steam to leave your boiler and head for your radiators. That steam is looking for a way *out*!

So it races up and out and into the pipes, pushing air ahead of itself because that's what's inside your pipes when the system first starts – air. And *lots* of it. The pipes are filled to the brim with air, and the steam can't go anywhere until that air gets out of its way. The steam has to plunge the air down the pipe and, hopefully, the air will cooperate by moving along peacefully. You see steam and air can't mix because they have different densities (steam is lighter than air). That's why steam rises when it leaves a boiling pot of water on your stove. It's like helium, lighter than air. The air scurries ahead of the steam and leaves the system through a large air vent, which should be somewhere near the end of the main pipe in your basement. This vent, the **main vent**, is the main exit door for the air.

If you're poking around your basement, look for something like this.

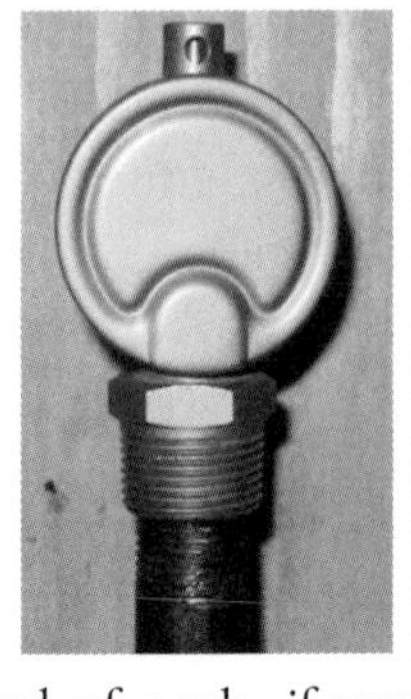

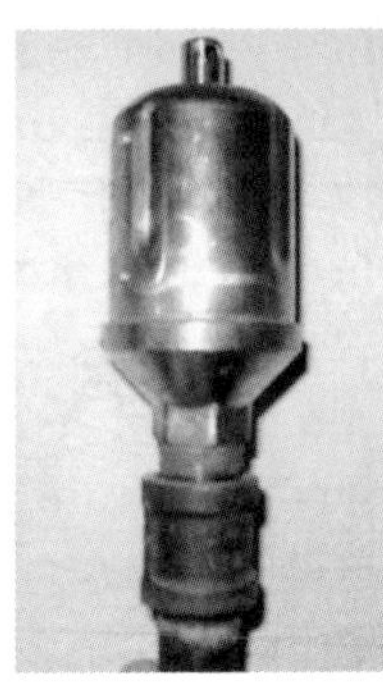

Follow that main pipe that runs from your boiler across the basement ceiling and around your house. There may be more than one of these main pipes, and more than one main vent. Take a walk to the end of each, if you do have more than one, and look for main vents. If you don't see them, look for a pipe plug in a screwed tee fitting near the end of your main. That's probably where the main vent once was. It failed and someone replaced it with a plug. You can't vent much air from a steel plug, but plugs cost less than main vents, and that's how this usually happens. Knuckleheads.

But let's get back to what *your* knucklehead is probably missing when he's cranking up the pressure on your boiler.

Here's what happens in an old steam system. We start out with pipes that are filled with air. The steam pushes the air from the air vent and then the vent closes (more about the mechanics of this later). The steam condenses back to water, and in doing so, it shrinks to 1/1700th the volume it occupied as steam. This shrinkage creates a vacuum within your steam piping. The air vents throughout the system respond to this (once they cool down a bit) by opening and letting the air back into the pipes.

See? Your steam system breathes in and out, just like you.

Now, we've got these steel pipes that are constantly getting wet (from the steam) and we have this air that's moving in and out of the pipes like the tide in a harbor. Steel plus Water plus Air equals what? Correct. Rust! And lots of it. Sure, it takes years for this to happen, but trust me, it happens.

Now imagine yourself as a hanging flake of rust inside a

pipe. The steam comes roaring down the pipe and yanks you from your dangling position and flings you down the line. Where are you going? You're going to follow the path of least resistance, aren't you? Sure you are. And that's what I'd do too. I'd head right toward that main air vent because that's where all the other guys (Mr. Steam and Mr. Air) are heading. And right behind me is more rust. And all sorts of other accumulated goop. Pretty soon, the inside of that main air vent is so clogged with crud that it can't close. Steam starts spewing out and someone picks up the phone and calls for help (or takes a ride down to the Home Center) and this is when the Laws of Economics overcome the Laws of Physics. Someone removes the leaking main vent and installs a pipe plug.

So where's the air going to go now?

It certainly can't get out of the pipes. The steam is pushing it, just as it always has, but the main air vent is gone. There's a pipe plug there now.

So what's the steam going to do?

Well, what would *you* do?

If someone (Mr. Steam) backed you (Mr. Air) into a corner (the end of the main) and kept shoving you, what would you do? Wouldn't you shove back?

Sure you would. And that's what compressed gases do. You squeeze them; they push back. The end result? The steam stops moving and the pressure builds to a point where it shuts off the burner. No fire, no heat. Simple.

And keep in mind that steam and air have different densities, so you can't mix them. Air trapped at the end of the main might as well be cement, as far as the steam is concerned.

Back to your knucklehead. He's standing on the outside, feeling the main pipe, and it seems like the steam has stopped right about . . . here. The pipe is hot right here, but cold over there, just a foot or so further down the line. Standing on the outside, it sure looks like there's something stuck in the pipe, doesn't it. There is. It's air. But your knucklehead's not thinking about air. He's thinking it's more a plumbing problem than a heating problem – like a sewage blockage. (I've actually seen Do-It-Yourselfers pour drain cleaner into their top-floor radiators in hopes that it will flow downhill and dissolve the "clog," which is nothing more than air. *Promise* me you won't do that.)

Anyway, your knucklehead figures that he can use steam pressure to blow free the "clog." So he raises the pressure and this only causes the burner to run longer, which wastes the fuel that you're paying for, while compressing the trapped air even more. The air gets mad and pushes back and the burner shuts off again. So your knucklehead raises the pressure even higher. He does this after walking along the main and feeling it get hotter further down the line. It gets hotter because the higher steam pressure is compressing the trapped air and jamming it further into that corner.

Your knucklehead winks at you. "We're making progress!" he says. He says this because he's a knucklehead.

Before long, he has the boiler pressure cranked up to a point where you could open a dry cleaners, and this is when your knucklehead shrugs and says, "That's about the best we can do. It's steam, you know. Wadda ya expect?"

And then he gives you a bill.

Which you pay.

It's steam. Wadda ya expect?

What your knucklehead doesn't know, and what I hope you're beginning to realize, is that you can move with a pinhole (or better yet, a main air vent) what you can't move with a ton of steam pressure.

Can you see it in your mind's eye? It's common sense, isn't it?

But it's not common practice.

Don't let anyone crank up your steam pressure. Look for those main vents. They can make or break a steam heating system. They're *that* important.

Dry steam *good.* Wet steam *bad!*

When you're thinking about steam you're probably thinking about the stuff that comes out of a teakettle on your stove. It whistles through that little hole and heads for the kitchen ceiling. That's how most folks see steam, but that's not good **dry steam**. You can't actually see dry steam. What you see coming from the teakettle is **wet steam**. It's tiny droplets of water that become visible because the steam is giving up its heat energy to the air. If you look very near the teakettle's hole you'll notice a clear space between the whistle and the visible "steam." What you're looking at (but can't see) is dry steam. That's the good stuff. Dry steam is as invisible as air.

And maybe you've heard the stories of sailors who work in the bowels of the ships that have the high-pressure boilers that run the turbines that spin the propellers that move the ships forward and backward. If those sailors suspect a steam leak they'll walk along with a corn broom held out in front of them. If the end of the broom suddenly gets cut in two, the sailor knows to stop walking. He has found the steam leak. It is as invisible as air, and as deadly as a laser.

Thankfully, we're not working with pressures that high but the good stuff – the dry, invisible steam – is what we're looking for when we do steam heating. We usually want it at *ounces* of pressure, though, not pounds of pressure.

Dry steam is a technical term that heating professionals and boiler manufacturers use to distinguish the good stuff from the bad stuff (the wet steam). Dry steam contains no more than two percent moisture. I'll tell you a lot more about this later on, but for now, just know that if the boiler water is dirty, or if the piping around the boiler isn't right, or if the burner isn't set up properly, the steam can get wet (and remember, that's a technical term), and this is bad news for you and your system. Wet steam won't heat your home properly. It will cause your fuel bills to be higher than they ought to be.

Put that thought on the back burner with the teakettle for a moment. We'll get back to it, but first, I have to tell you a story.

What you may not know about temperature scales

Most of the scientific concepts we take for granted nowadays started as an idea that some person dreamt up. Take temperature, for instance. Water boils at 212 degrees Fahrenheit at sea level. You learned that when you were a kid. But have you ever wondered why the number is 212, and not, say, 210 or 215?

I did.

So I went and found out.

Here's the deal. Before we had thermometers no one knew how hot or cold it was outside. The concept of temperature measurement didn't exist. No one thought about it. Tough to imagine, isn't it? Way back in the day you'd go out into the July heat and say, "Boy, it's sure hot today! It must be, what . . . " And there you'd get stuck because the folks who invented temperature measurement hadn't yet done so. So you couldn't say, "Man, it must be 98 in the shade today!" or "I hear it goes up to 120 in Phoenix. But it's a *dry* heat." Nope, in those days, all you could do was mop your brow and complain. There were no thermometers. There was no concept of temperature measurement.

Stop for a minute and think about that. We base so much of what we do from day to day on the temperatures inside and outside our buildings. We're forever fiddling with thermostats, trying to find that perfect level of comfort. Some people like it at 72 degrees Fahrenheit, while others prefer 65 degrees Fahrenheit, and if there are more than two people in the room chances are the temperature will *never* be right. It sure is strange to think of a time before we had Fahrenheit and Celsius, isn't it?

Speaking of which, I get very confused when I'm in Canada and Europe and the woman on the TV tells me it's going to be a lovely 25 degrees today. I'm reaching for my overcoat until I realize she's talking Celsius and not Fahrenheit.

And actually, she's not even talking Celsius (even though most of us say that) because no one uses the Celsius scale nowadays. Anders Celsius was the guy who dreamt up that one, and he did it a long time ago (1742). Mr. C. decided to make the boiling point of water zero degrees and the freezing point of water 100 degrees. Did you know that? He could get away with this because he was Anders Celsius, and who could argue with Anders Celsius when it came to the Celsius scale. He owned the darn thing.

But no one uses Celsius anymore; we use the Centigrade scale but Celsius usually gets all the credit, even though he had nothing to do with turning the scale upside-down to get to Centigrade. Doesn't seem fair, does it?

In the United States, most of us are still swearing by the wisdom of Herr Gabriel Fahrenheit. Gabe was a German merchant and the first guy to make a mercury thermometer. This was also a long time ago (1721). He had to come up with a scale to go along with his new thermometer, and he needed to have fixed points on that scale, so this is what he did. He used as zero degrees Fahrenheit the temperature of the coldest stuff he could imagine in 1721, a mixture of salt and a chemical called sal ammoniac. The other fixed point was to be the normal temperature of the human body, which he called 24 degrees Fahrenheit. Did you know that? He could do this because he was Gabriel Fahrenheit. And who's going to argue with this guy? It's *his* scale, for Pete's sake!

Anyway, on this original Fahrenheit scale, water freezes at eight degrees Fahrenheit and boils at precisely 53 degrees Fahrenheit. Gabe took his thermometer around and showed it to people who had no concept of temperature measurement.

They had never needed a thermometer, had never heard of temperature measurement, and here was this guy, Gabriel Fahrenheit, trying to sell them both.

And you think you've got it tough?

But he did it, didn't he? He sure did! But not without going off on a few tangents. You see some people mentioned to Gabe that the mercury moved past those numbers on the scale pretty quickly, and wasn't there a way he could slow down the mercury so that they could get a better sense of what was going on.

Well, he couldn't slow the mercury, of course, so he did the next best thing. He changed the scale by multiplying the entire thing by four. Water, which previously had frozen at eight degrees Fahrenheit now froze at 32 degrees Fahrenheit (isn't that marvelous?). And instead of boiling at 53 degrees Fahrenheit, water would now boil at 212, because he said so. Hot stuff, eh?

And by the way, the normal temperature of the human body is actually 96 degrees Fahrenheit (24 X 4 = 96) because he said so, but we've now all agreed to call it 98.6 because Gabe is currently on the other side of the lawn and not in a position to argue with us.

You can't make this stuff up.

The best part of all of this (at least for me) is that this guy Fahrenheit just woke up one day and made all of this up. Today we take it as gospel, like some great scientific truth, written on granite, a force of nature, something you could take to the bank. Absolute truth!

Truth is, though, Gabe could have called it anyway he saw it and we would have gone along with him because he was Gabriel Fahrenheit and this was *his* scale.

Hey, what the heck.

And you can do the same if you're willing to put in the hours and work as hard at it as he did. Go ahead; establish the Schwartz Scale, the Murphy Meter, Ropinski's Range or anything else your fertile mind can conjure. Go ahead; knock yourself out. Get out there and tell the world!

Science will remember you forever.

Think of it. What makes a foot a foot, a yard a yard, a meter a meter? Someone just got up one day and conjured. Then he or she worked really hard, and convinced the rest of us to just shut up and go along. Why is a troy pound different from an avoirdupois pound? And isn't it amazing that people throughout the world can get any engineering at all done with all these different terms?

There was a guy named Delisle who was also into thermometers and temperature measurement. He introduced the Delisle scale in 1724. Ever hear of it? He was following dead on the heels of Gabe Fahrenheit and he, like Celsius, decided to call the boiling point of water zero degrees. He also figured that 100 degrees should be the temperature of a cellar in the Paris Observatory.

But on what day?

And in which corner of the cellar?

Or maybe it was in the middle.

Didn't matter; he was making it all up and it could have been anything and anywhere as long as it pleased him. And in spite of that wacky decision (the basement of the Paris Observatory?), this became the scale that Russia chose to use for many years until they wised up and switched to the Reaumur scale.

Ever hear of him?

René Antoine Ferchault de Reaumur sold alcohol (not vodka) thermometers, and the boiling point of water on the Reaumur thermometer is 80 degrees. Much of France still uses this scale. Go figure.

Isn't it amazing that we all managed to get where we've gotten. We can't even agree with exactly how hot or cold it is, inside our homes or out in the street.

But I digress. We'll get back to your steam system in a second, but I just *know* you're wondering who invented that fabulous Centigrade scale, which most of the world agrees is the most sensible (no pun intended) scale of all? Well, that marvelous feat goes to Carl Linnaeus, a Swedish botanist who also established the modern binomial system (genus plus species) for naming plants and animals.

The Centigrade scale was just something he came up with in his spare time. And isn't that wonderful?

See all the things you can learn when you become curious about your old steam heating system?

Onward!

A watched pot never boils

Now that you know far more than you'll ever need to know about temperature, it's time for us to talk about heat, and to discuss that old saw mentioned above. The watched pot. It never . . . seems . . . to . . . boil.

Actually, it does. You just have to watch it long enough.

Okay, basic question. What makes the water boil in a pot?

Heat, right?

But how much heat does it take to make water boil? That's what you want to know when you've got a steam heating system in your home because that's what you're paying for when you heat your house. You're paying for the *heat*. Which brings me to what I hope will be a plain-English explanation of **heat** and how we measure it. And please put away your thermometers at this point in our journey because, when it comes to the making of steam, thermometers really aren't that useful.

Okay, we'll begin with a pound of water (which is about a pint). Now as you've known since you were a pup, you can have that pint of water three ways. You can have it as solid ice in your freezer. You can have it as a liquid pouring from the tap. Or you can have it as vapor. When there's heat involved (as with a teakettle on a hot stove, for instance), we call that vapor **steam**, but there's another type of vapor, and that's the kind you get when water e***vapor***ates (get it?). This doesn't require any additional heat; it just happens on its own.

Okay, now there's an odd thing about water. It can be either a solid or a liquid at 32 degrees Fahrenheit, and it can also be either a liquid or a gas at 212 degrees Fahrenheit. What causes the change in state (from solid to liquid or from liquid to gas) is heat. And it's a type of heat that we can't measure so the Engineers of Yore decided to call this **latent heat**. It's heat that's there, but you can't measure it with a thermometer. If you have steam heat, this is the stuff that's going to heat your home. Latent heat.

The other type of heat, **sensible heat**, is easier to imagine because you *can* measure it (or *sense* it, as in *sensible*) with a thermometer. Sensible heat is the heat that we add to a pound of liquid water to bring it from 32 degrees Fahrenheit to 212 degrees Fahrenheit. This water is not ice at the beginning of its journey or steam at the end. It's liquid all along the way.

And that brings me to a term you should get to know – **British thermal unit**.

A British thermal unit is the amount of heat required to raise one pound of water one degree Fahrenheit.

What we have here is a term that Thomas Tredgold, an Englishman, came up with shortly after the British had kicked the snot out of Napoleon. Mr. T. figured that since they were now sitting on top of the world, he and his countrymen had earned the right to coin engineering terms that were as English as a bowler hat. And why not?

Thus, it's a **B**tu (*British* thermal unit) and not a **P**tu (*Polish* thermal unit). So there.

Anyway, let us now sense the sensible heat. We begin by pouring a pint of 32°F water (drip it from some ice cubes) into a pot on the stove. Add one Fahrenheit scale thermometer. And now add heat. One Btu will raise the temperature of this pound of water one degree Fahrenheit. So, to get from 32°F to 212°F we'll have to add the difference between those two numbers (212 minus 32), or 180 Btus.

That's sensible heat.

Now, what we have at this point is 212°F liquid water. This is **not** steam. This is very hot, liquid water. To get the liquid to turn to steam we have to add more Btus. A *lot* more. In fact, we have to add an additional 970 Btus to that one pint of water before it will turn to steam. That's more than *five* times the amount of heat it took to bring the liquid water from 32°F to 212°F, and that's the stuff that's going to heat your home because when the steam condenses inside your radiators, you'll get nearly all of that latent heat energy back (some gets lost along the way and I'll tell you about where it goes in just a little while).

Keep in mind, too, that this is with no pressure at all. That pint of water will contain a grand total of 1,150 Btus (that's both sensible and latent heat) and this is why you don't need very much steam pressure to heat your home. It's also why your knucklehead is mistaken when he tells you that you have to crank the pressure higher than two-psi to heat the bedrooms on the second floor.

The only reason you need any pressure is because the steam has to overcome the friction that the insides of the pipes present as the steam soars between your boiler and your radiators. And since the pipes are large (thanks to the Engineers of Yore) there won't be that much friction to overcome.

So crank down that pressure. You'll save money on fuel, and your home will heat better. And if you find that it doesn't, it's probably because the air can't get out of the pipes quickly enough. Go look for those main vents. I'll bet you find pipe plugs instead of main vents. Economics vs. Physics.

Are you starting to get it? It's really not that complicated, is it?

You just have to think like steam!

It's chilly out there!

Sure *is* chilly out there! Out there in the system I mean. And I'm talking about how the steam feels after it leaves your boiler. It has all that pent-up latent heat – heat that you paid good money to produce – and the first thing the steam meets is a network of cold steel. All the pipes in your system are as cool as the air around them, and the steam faces a mighty task in having to bring all that heavy steel up to steam temperature before it can work its way up to your radiators. And that's why the Dead Men wrapped their pipes with insulation. They wanted to keep them tucked in so that the steam could hang onto as much latent heat energy as possible as it pushed the air out of the system and fought its way to your radiators.

Unfortunately, the Dead Men used asbestos, which was perfectly fine back in the day (who knew?), but you've probably gotten rid of that nasty stuff already, right? Or if not, you're probably thinking about getting rid of it. Aren't you? If it's torn, worn and going airborne, you *should* be thinking about getting rid of it. I would.

But now think like steam. If the pipes in the basement aren't insulated, where are you, as steam, going to condense and give up that massive amount of latent heat energy? Would you condense in those cold basement pipes downstairs? Or would you condense in the radiators upstairs?

Downstairs, right?

And keep in mind there's a limited amount of steam to work with here. Remove the asbestos and don't replace it with another type of insulation and you'll wind up with a very cozy basement, but very chilly rooms upstairs.

There's only so much steam available to you. It's like an ATM. You can't take out more than you put in.

And raising the pressure won't help because raising the pressure doesn't make the fire inside your boiler any larger; it just causes the boiler to run longer, and that costs you money.

We'll talk more about this later, but for now, just know

that one of the most common causes of steam heating problems (banging pipes, spitting air vents, high fuel bills, uneven heat) is the lack of pipe insulation.

Got that?

Good!

Now back to the steam. It's expanding like crazy as it races out of your boiler and into your pipes. And even if your pipes are well insulated some of the steam is still going to condense along the way to your radiators. There's no getting around that and this will continue to happen until all the pipes reach the normal steam temperature that you find in a heating system (which is about 215°F).

Steam is a gas that wants to be a liquid. It's as plain as that. It will give up its latent heat energy to anything it meets that is cooler than it is. It will always travel toward the air vents, following the path of least resistance, which is why the Dead Men took such care in sizing their pipes, and why you shouldn't change any of those pipes without first thinking about what will happen afterwards.

The steam may not go the way you want it to go. Steam is like that. It sticks up for the Dead Men. Trust me on this.

Keep in mind, too, that once the steam starts to condense we're going to have this other stuff that we call **condensate**, which is just a fancy way of saying **water that was once steam but has now condensed**. The condensate is going to have to get out of the steam's way and that's where pipe size and pipe pitch and insulation become so important.

If the pipe is the wrong size, the steam will pick up the water and fling it down the line with ear-splitting violence. We call this **water hammer**. You probably know this sound. It may even have been your inspiration for buying this book. Good.

If the pipe is the wrong pitch, the water won't be able to flow by gravity along with the steam (or, in some cases, in the opposite direction).

If the pipes aren't insulated, you're going to have a lot

more condensate than the Dead Man who designed your system intended for you to have at any given moment, and this can also lead to water hammer.

So please don't go moving pipes on your own. Check with a pro that knows steam. I'll tell you how to find one of those later on in the book. For now, just believe me when I say that *everything* in a steam system is where it is for a good reason. You can't just start lifting pipes and expect things to work.

If all goes well, the steam condenses and then flows by gravity back to your boiler where it will once again become steam. It's the great circle of H_2O life. And it really is pretty simple. We begin as liquid, turn to steam, chase the air out of our way, turn back into a liquid and roll downhill and back into the boiler, and as we do, we suck the air back into the pipes and radiators. Just a big wonderful circle.

And if all is well within your system, you should expect your rooms to be cozy within 20-minutes from the time your boiler first starts. That's what I would consider normal in a typical house. And all of this should happen without any noticeable noise. Your pipes should not bang. You should not hear air hissing from the air vents on your radiators. It should all be very peaceful and wonderfully comfortable.

I know that's not the way it is in your home, but that's the way it *should* be. And that's where we're heading. Together.

Next stop?

What makes the steam go *that-away?*

When your boiler turns water to steam, it causes that sudden and massive 1700:1 expansion that drives the newborn steam out of your boiler and into your steam pipes. The Dead Man who designed your steam system sized the pipes so that the steam would move at an average speed of about 30 miles per hour. In some systems it can go even faster than that.

So why don't your radiators get hot instantly? It's because the air is standing in the way of the steam. The air has to leave the pipes and radiators on every heating cycle, which is why I keep harping about the importance of proper air venting. The steam will move a *lot* faster than you can run, but only if the air gets out of its way.

Now, I have to tell you something about steam and how pressure affects its speed. This is probably going to seem pretty bizarre (it did to me when I was first learning about steam), but it's true. Ready? Here it comes.

When you raise the pressure on your steam boiler, the steam doesn't go faster; it goes *slower.*

Seems crazy, doesn't it? It goes against logic. You would think that if you raised the pressure at the boiler you'd be pushing the steam harder, so it should get to your radiators faster, right?

It doesn't. And here's why.

Take a look at the nameplate on your boiler and you'll find a rating in **Btuh**. As I told you earlier, that's short for **B**ritish **t**hermal **u**nits per **h**our. In other words, your boiler can raise so many pounds of water one degree Fahrenheit over the course of one hour. The size of the fire inside your boiler determines this. You can't get more heat out of your boiler than the fire puts in (just like the money in the ATM). If you raise the steam pressure on your boiler the amount of heat coming out of that fire won't increase or decrease by a single Btu. You simply can't get more out than what you put in, and what you put in is what the nameplate states.

So in one hour, your boiler will produce a very definite amount of steam. It won't do this in 59 minutes, or in 61 minutes. It will do it in exactly one hour. That's important, as you'll see.

Okay, now think about the pipes through which the steam will be traveling. The size of those pipes is fixed, isn't it? I mean no one is going to show up and change the size of the pipes when the boiler starts to make steam, are they? Of course not! The steam will enter those pipes and it will flow toward the air vents, and it will lose pressure along the way due to friction (caused by the steam rubbing against the insides of the pipes), and the steam will also lose pressure as it condenses.

So we have a fixed amount of steam traveling within pipes that have a fixed amount of internal space. With me so far? Good!

Okay, let's consider the steam, which is a gas. When you compress a gas it takes up less space, doesn't it? Sure it does. You know this from everyday life. Just think about helium for a moment. You go to the store to buy a bunch of balloons for a birthday party. When you get there, you see that other people are also having birthdays. The ceiling at the party store is thick with bouquets of balloons, all waiting to be picked up by happy people.

Take a moment to notice how much space those balloons are taking up as they press against the ceiling in the party store. Now look at that tall helium canister that the clerk is using to fill all those balloons. See how small it is compared to the volume of all those balloons? That one helium tank at the party store will fill about 600 balloons, each 11 inches in diameter. But how do they get that much helium inside that steel tank?

They pressurize it, of course (which is why the tank is made of steel), and when you open that tank the helium comes roaring out and expands like mad. Gases do that when you let them loose. They expand and contract in relation to the pressure that you apply to them.

Okay, back to your steam system. I told you that the steam wants to move at 30 miles per hour, and that's based on it leaving the boiler at a pressure of just a few ounces per square inch (which is what you have when your boiler first starts making steam). Now one pound of water will fill a single pint glass. One pound of steam at zero pounds per square inch will fill 27 cubic feet (that's a cubic yard, three feet on each side). Raise the boiler pressure to 10 psi and that same pound of steam will take up just 16 cubic feet. The higher you raise the pressure on the boiler, the more compressed the steam will become. It's like the helium in the tank at the party store.

Okay, hold that thought for a moment. I want you to take a ride with me. I want you to think about 1,000 Volkswagen Beetles. Imagine them in a rainbow of colors and put a driver in each one. Let's call each driver Mr. Btu.

Now imagine a one-lane road that's one mile long. It can be a country road or a highway; it's your choice, but what you're going to do in your imagination is move those 1,000 Volkswagen Beetles down that mile-long road in precisely one hour. That's the goal – to move 1,000 Mr. Btus per hour. You have to have that last car crossing the finish line when the stopwatch hits the one-hour mark. You can't start a moment sooner, and you can't finish a second later.

This is what we mean when we talk about Btuh (British thermal units per **hour**). If I say 1,000 Btu**h**, I'm not talking about 1,000 Btus in 59 minutes, right? Nope, I'm talking about that much heat in one *hour.*

So when comparing the movement of steam to the movement of Volkswagen Beetles (something you do every day, I'm sure), please keep time in mind.

Let's see. What else? Oh, did I mention that those Volkswagen Beetles are under 10-psi pressure? Well, they are. Add that to your imagination. You've got the Beetles moving down the road (or through a steam main) and they're all under 10 pounds per square inch pressure, which is why they're so small.

Okay, time to get rid of the Volkswagens. We'll release the pressure on them, and when we do that, they'll expand like a gas. In fact, they'll magically transform themselves into big yellow school buses (work with me on this).

Now, bring the buses around to the starting point. Who's driving? The 1,000 Mr. Btus – the same guys who were driving the Volkswagens. Your job is to be in charge of the team that's going to move those 1,000 big yellow school buses down the same road that the Volkswagen Beetles just traveled. And you have exactly one hour to do this. You can't take a second more or a moment less; you have to do it in *exactly* one hour.

Ready for the big question? Which vehicles have to travel faster to get the job done – the Volkswagen Beetles, or the big yellow school buses? The drivers are the same. The road is the same. The time is the same. The only thing we've changed is the size of the vehicles. So, which has to go faster to cover the distance in exactly one hour?

The school buses have to go faster, right? Of course, they do! They have to go faster because they're bigger. Just think of how much longer a line of 1,000 school buses would be, compared to a line of 1,000 Volkswagen Beetles. If you have to get those two lines down that mile-long road in *exactly* one hour, the busses will naturally have to go faster than the Beetles. Make sense?

That's why low-pressure steam moves faster than high-pressure steam. It's bigger.

Think about it:

✔ If the quantity of steam the boiler produces in one hour is fixed (which it is).

✔ And if the pipes will always have the same amount of internal space (which they will).

✔ And if that steam takes up more space when it's at low pressure than it does when it's at high pressure (which it does!).

✓ Then we can say with absolute certainty that our given amount of steam, over the course of that one hour, within those pipes of a fixed size will *definitely* move faster at low pressure than it will at high pressure. It has to. At low pressure, the steam is as big as a school bus, and it's turbocharged with latent heat.

The Dead Man who designed your steam system knew this. He designed the pipes to deliver steam to your radiators at very low pressure. The lower the pressure, the faster the steam will travel (and the safer and more economical the system will be). If the steam isn't getting to your radiators, it's not because of a lack of steam pressure. Something else is going on.

Which brings me to how steam often seems to have a mind of its own. Why does it sometimes go this way and not that way (like to your cold bedroom)? It doesn't know up from down. It's a gas and all it knows is out. It's looking for the air vents, and that's why it goes the way it goes. It goes toward the air vents. There has never been a steam heating system that didn't need to be vented in some way. Sometimes, the air vents aren't apparent, but believe me, they're out there, and they're essential to the system's operation.

The steam is also following the path of least resistance. The Dead Man who sized your pipes probably did his best to balance the long runs of pipe against the shorter runs of pipe by selecting pipes of different diameters. The larger the pipe, the less resistance it offers to the steam. If he didn't do a good job (which was rare during those days) then the system won't balance well. You can easily mess up the Dead Man's good work by reducing the size of the steam pipes in your home. Looking for trouble? Do *that.*

Removing insulation can also affect where the steam goes because when steam enters an uninsulated main it condenses more quickly than it will in an insulated main. This sudden collapse of the steam can create a partial vacuum in parts of the system and the steam will just naturally rush that way to fill the void (while perhaps avoiding that cold bedroom of yours).

Steam really doesn't have a mind of its own; it just seems that way sometimes. Steam goes where it can, and where it finds it easiest to go, and these are reasons why it does this. Steam won't necessarily go to a chilly room in your house even if you shut off the radiators in the rooms that are too hot. The key to getting the steam to go where you want it to go lies mainly with keeping the pressure low, making sure the pipes are insulated and having those all-important air vents in all the right places. More on this as we move along, but first, let's talk a bit about those strange noises you're hearing. Are they goblins? Nope, just the steam (and condensate) complaining because something's amiss in your system.

Let's see what that is.

Things that go bump in the night (and during the day!)

The noises that your steam system is probably making aren't normal. The banging and clanging, the hissing and spitting – all *not* normal.

What I'd like you to do is think of your system as an unhappy child (one weighing several tons). This child is just trying to get your attention, and this child, as you know, can be very persistent. And quite loud.

I learned about banging radiators when I was a lad in New York City. We had steam heat in our apartment and the radiators banged and ticked and complained most of the time. We thought that just meant that the heat was on. Sometimes, my father would bang on the pipes with a claw hammer, just to let the superintendent know that he knew that the heat was on. The super would bang back, and my father would bang some more, louder this time. And then it was the super's turn. And then my father, who was a bit of a maniac, would take a walk down to the super's apartment. And we all thought that this was quite normal.

Which it's not.

The reason I know for certain that it's not normal is because I have all these old books that tell me so, and I once had an experience in New York City when I was a younger man, and that experience opened my eyes to what's possible.

It went like this.

Old Buildings, New Steam

The call came in on a day when a merciless wind whipped out of the Northwest, as it so often does during a New York City winter. The call was about a building with no heat.

Since the steam heating system in the old, but newly renovated, apartment building was under warranty, the folks at New York City's Department of Housing Preservation and Development passed the call on to the installing contractor, and he, being a reputable businessman, immediately dispatched a serviceman to get the burner started.

There was a problem, though. When the serviceman got to the basement and looked at the boiler, he noticed right away that the burner was missing.

Now, even the best serviceman in New York can't fix what's not there, so this presented a genuine challenge to the guy. He was standing there, scratching his head and wondering what to do next, when a young man arrived at the boiler room door with the perfect solution. This young man expressed his sympathy for the serviceman's dilemma, and then he explained that his brother sold burners that were absolutely guaranteed to fit that very boiler, and that he, the young man, could get one for the serviceman for just 50 bucks. And he could do it right now. No waiting.

Well, this being New York City and all, it seemed like a pretty good deal to the serviceman. He called his boss and got authorization for the (cash only) purchase. The young man then went away for a few minutes and returned with a little red wagon in which sat a rather large burner. And as promised, it fit perfectly. How about that!

You can't make this stuff up.

New York City owns a lot of buildings. It picked them up one at a time. Most of these buildings came to the City by way of landlords who decided to stop paying their taxes. The City calls this "in rem housing." In rem is a Latin legal term, meaning "against a thing" – the "thing" being you if you don't pay your taxes. Ya snooze, ya lose, pal.

Back in the Eighties, many of the in rem houses were visible from New York's major highways, and this was doing nothing for tourism, so the folks in charge came up with an idea. They decorated the blown-out windows of these tenements with hand-painted sheets of plywood. Each window had a festively colored set of curtains and a flowerpot or two. Brilliant. It was like putting shiny pennies on the eyes of a corpse.

By 1984, the homeless situation was getting desperate. A group of City agency people began looking at both the homeless and the in rem housing situation as one problem. They reasoned that they might be able to fix the abandoned buildings and create housing for the homeless, and this just might solve both problems.

They began with a patch-it-and-get-out pilot program. Their plan was to do as much as possible for as little expense as possible. Unfortunately, most of the buildings had deteriorated to a point where there weren't enough places to stick the patches. Before long, the agency people realized that it was actually cheaper to gut those old tenement buildings and begin anew from the shells. Most of the people on government assistance were women with small children. The City had been housing these people in single-room-occupancy hotels at a cost of about $34,000 per family, per year. The cost of renovating one of the abandoned apartments was about $65,000. Gut-rehabilitation of these buildings made sense since the payback period was less than two years. Which was why the City was moving in that direction, but it was happening slowly.

Then, an important court decision came down. It ruled that a single-room occupancy hotel was no place to raise children, and this pretty much made the decision for the City officials. Women and children were being tossed out of the single-room occupancy hotels with nowhere to go except into the in rem tenements, and that's when everything started to move fast.

There were hundreds of these buildings, and most were five or six stories tall. They all went under reconstruction at the same time and the heating system that the City chose to use (and these were brand-new heating systems) was one-pipe steam.

Surprised?

Don't be; it made sense. First, steam once heated these tenement buildings. The City figured that if it worked once, it should work again. Second, when the heat goes off (in other words, the next time the burner goes for a ride in a little red wagon) the pipes won't freeze because, in a steam system, most of the pipes hold no water – just air and steam.

Next, if a pipe breaks in a steam system, there's very little damage. Not so with a hot water system where you can have quite a flood.

Another good point: Steam systems are pretty easy to design and to balance (if you know what you're doing).

And steam systems are rugged. They can take a beating and run for years. That's why there are still so many of them left in America. And the parts of the system that the tenants would have access to would be difficult to damage.

Steam heating also presents no static-pressure problems. This is a concern in high-rise construction that uses hot water heat. The higher you stack water, the more pressure you get down there at the bottom floors. That often calls for special (and more costly) equipment. There are no static pressure problems with steam. Another plus!

But the most important reason that they chose to go with steam heat (and you can't make this stuff up) is that, with steam, there's nothing worth stealing from the jobsite. This is not true with hot water heat. Hot water systems have copper pipes and copper fittings and copper radiators and brass valves and these metals make scrap dealers drool.

Steam pipes, on the other hand, are made of steel, and the heating units in the apartments are also made of steel, and steel is not worth toting out of the building. The scrap dealer would just laugh at you. Make sense? It did to them! At the time, the City's Project Development Coordinator said, "A contractor can look at a truckload of copper tubing and see a truckload of copper tubing, but to a drug addict, that truck looks like a jewelry store." He went on to speak from personal experience about this. "When we tried a hot water heating system in one of these buildings," he told me, "the only way we could keep the copper in the building beyond the first day was to paint it black as soon as it arrived on the jobsite. Once it was black, the locals thought it was steel and left it alone. If it looked like copper, it didn't make it through the night."

Unique New York.

So to the folks in charge of this massive rebuilding project, one-pipe steam looked like a mighty tasty solution. First, it's a simple system. Second, it's not worth ripping-off. Third, if someone should cut a pipe once the system is up and running, there won't be a flood. Instead, that person will get burned. Which is exactly what that person deserves. So there.

The problem they faced, however, was that most of the City engineers had no idea how to design a one-pipe steam heating system. No one had done this for at least 40 years, and none of the current engineers had been around back then.

So they hired me.

Now I'm not an engineer, and I've never worked in the trade, but I had this wonderful collection of antique engineering books, and I had a lot of hours of poking around old buildings with a lot of old-timers, and I do know how to explain this stuff in plain English. I guess the City figured I'd do in a pinch.

So we all got together for a few days and I spoke to them of all the things that I'm telling you, but on a deeper level because they were actually going to *design* these steam heating systems. They had to know about how to properly size the pipes, and how much to pitch them so that the steam and the water got out of each other's way. They needed to know about system balance and air venting and how to get the condensate back to the boiler once the steam reached the radiators. They had to be able to work with the existing geometry of the buildings, and they had to know what to do if things didn't go according to the textbook. They also needed to know where the limits were, and to not push those limits. And most important, they needed to overcome that nagging feeling that there's something mystical about steam heat, and that you can't possibly understand it unless you're 90 years old.

And some time went by after our classes, and then they issued the plans and specifications to the bidding contractors, and they gave those folks a lot of latitude on the pipe sizing, knowing that the contractors would have to deal with the real-world conditions of the existing buildings. If a pipe couldn't go where the plan said it should go, the contractor would have to re-route it and that often called for a change in pipe size. So the contractors had plenty of leeway with these buildings and that worked out to be a good thing. They had the freedom to change things. The bottom line was that they had to deliver turnkey systems that didn't bang or spit water from the air vents. That was the thing. No banging. No spitting. Imagine that with steam.

And when the jobs were done I went to see them and I tell you they were things of beauty. The steam moved through the basement pipes faster than I could run. It shot up the risers and into the apartments quicker than I could gambol up the stairs. It was *that* fast. And there was no noise.

And I learned that this is the way steam is supposed to be.

So let's see if we can get your system working that quickly and that quietly. Read on.

The noises you might be hearing (and what causes them)

Let's begin with the **hissing**. It's probably coming from the air vents on your radiators. It's not an unpleasant sound, but it's also not a normal one. The hisssssssssssssssss that often accompanies the start of a heating cycle is the sound of the air moving across the small holes in you radiators' air vents. Push air through a little hole fast enough and you're going to get noise. The harder you push, the louder the noise. Your air vents are like tiny musical (or maybe not so musical) instruments.

But consider this. If the air had more ways to get out of your system, it wouldn't have to push as hard at any one of the holes (air vents) in the system, would it? Of course not. Just think about a crowd leaving an arena after a sporting event. The more exits they have available, the less pressure there will be at any one of those exits. Open all the doors and people will leave the arena in an orderly way. Lock most of the doors and watch what happens. Everyone starts pushing and things get crazy.

The air in a steam system works in a similar way. At the start of each cycle, all the pipes above the boiler's waterline, as well as all the radiators, are filled with air. We know that the steam has to push that air out the air vents. If there are just a few air vents, the air is going to rush from them and make a hissing noise. Add more vents and the system gets quieter. Make sense?

And once again, this is why those main vents near the ends of the big pipes in your basement are so important. Not only do they help balance the system, they also get rid of the majority of the air that's in the piping so that it doesn't have to leave the system through your radiator air vents. And that means you won't have to listen to it upstairs.

And think about this. There are those flakes of rust inside your pipes that will always be there because the system is constantly corroding. The faster the air moves past those flakes, the more likely it will be that the flakes will wind up inside your air vents. And when a radiator air vent gets clogged shut, not only will that radiator not heat properly, but the other vents throughout your system will now have to pick up the slack. They'll be venting even more air, and that air will be moving faster than it should. So these vents will be making more noise. And the faster the vents vent, the better the chances are that they'll also get clogged. The more vents you lose, the worse it gets for those remaining.

So when you hear an air vent venting, this is not a good sound. This is the sound of the vent crying out to you for help. It's working too hard, and it needs your attention. If you don't help it, it will die. And that will cost you money.

So, make sure your main vents, and all your radiator vents, are clear of debris. Wait for a day when the system is off and cool to the touch. Use a wrench to remove the air vents and see if you can blow air through them. If they're clogged with debris, you won't be able to do this. I'll tell you later how to clean air vents.

The next noise you might hear is **ticking**. This is the sound that pipes make when they expand and contract. Ticking is especially noticeable if a pipe comes in contact with the wall or the floor as it passes through on its way to the next room. The Dead Men knew how to hang pipes to minimize expansion and contraction noises, and they also knew that the holes through which pipes pass must be wide enough to allow the pipe to thicken as it heats without grabbing the wood and lifting it.

Ticking sounds will happen (if they do happen) almost always at the beginning and end of a heating cycle, and they can be tricky because the sound carries well through solid pipe. If you're hearing these noises, look carefully at the places where the pipe touches the building. If the holes are too small (which often happens when a floor is replaced), widen them.

In older apartment buildings, and here I'm thinking about buildings that are five or six stories tall, the Dead Men would use a piping escutcheon to anchor the steam riser that went up through the building. Typically, they'd place this escutcheon against the floor and ceiling of the third story (in a five-story building). The escutcheon screwed onto the riser and held it tightly in place, forcing it to expand in both directions, away from that anchored point in the middle. If remodeling has led to the disappearance of that escutcheon, ticking noises may follow. If you live in an apartment building, you all share the same heating system. One tenant's decorating idea can often become another tenant's noise problem.

And that brings me to the loudest noisemaker of them all – **water hammer**. This is the sound you get when steam that's moving at 30 or more miles per hour picks up a slug of water and fires it down the pipe like an artillery shell leaving a cannon. The water, which is not compressible, meets a turn in the pipe and what you hear is the collision. CLANG!

Another type of water hammer occurs when there's too much water in the pipes (usually because of missing pipe insulation or poor pitch on the pipes). The steam roars over the top of the water, creating waves, just like a high wind does when it passes over the ocean. The waves crest, touching both the bottom and top of the pipe, and in the furrow between the waves we have a trapped pocket of steam.

That steam suddenly condenses and shrivels into nearly 1/1,700th the space it occupied as a gas. That creates an instant vacuum between the waves. The waves immediately rush in to fill the void, crashing violently into each other as they do so, and you get to listen to this while trying to relax on your couch. It sounds like a madman hitting an anvil with a 10-pound sledge.

Or my father fighting with the building superintendent.

There are a bunch of things that can cause water hammer. I'll take you through them one at a time.

First, there's **wet steam**. You'll recall this is the technical term that we use to describe steam that contains more than two percent water by volume. It's not something that you can test for in your home (testing requires expensive laboratory instruments), but it usually shows up as violent action in your boiler's gauge glass. Whenever I suspect wet steam is present, I round up the usual suspects:

- ➤ Dirt or oil in the water
  (which causes the water to boil violently)
- ➤ A bad fire in the boiler (which does the same)
- ➤ Incorrect piping around the boiler
  (which pulls water from the boiler)
- ➤ A boiler waterline that's too high
  (which tosses water into the pipes)

Wet steam allows too much water to get into the system piping, and when that excess water meets the steam, you'll often get water hammer. It usually requires the services of a pro to get to the bottom of this one.

Here's another common cause of water hammer. If the steam pressure is too high, the condensate often has a tough time making its way back to the boiler. That leaves too much water up in the pipes and the steam starts knocking it around. Crank down that pressure!

And then there are bad steam traps. I'll tell you more about these later when we talk about two-pipe steam systems, but for now, just know that steam traps do just that – they trap steam. They're automatic on/off valves that keep the steam from going into places where it doesn't belong. If your system has bad traps, the steam will get into the return lines and cause them to hammer.

Partially plugged air vents can also lead to water hammer because they can hold back water that's trying to return to your boiler. Hold your finger over the top of a drinking straw and lift it from a glass of water. Notice how the water stays in the straw? That's because you're blocking the air vent at the top of the straw. Good air vents help you avoid bad noises.

And I know I mentioned this before but all those pipes *really* need to be tucked in with a blanket of insulation. Without that blanket, the steam will condense very quickly and dump water in the places where the steam needs to be. Too much water mixed with steam gives you a racket in your pipes.

And then there's the pitch of the pipes and the pitch of the radiators. Steam pipes need to pitch downhill in the direction of travel at least one inch for each 20 feet of length. And that's when the steam and the condensate are moving in the same direction. If a steam pipe is back-pitched (where the steam and condensate are going in opposite directions) the pipe's pitch has to be twice that – one inch for every 10 feet of travel. And that pitch has to be continuous. If there are sags in the pipes, water will gather at that low point whenever the heat is off, much like water gathers in poorly pitched rain gutters. The next time the steam enters those pipes it will pick up the water and slam it down the line. BAM!

Another common cause of water hammer is a **wet return** line that clogs. A wet return is any line that's below the boiler's waterline. It's a natural settling place for all the crud that's in a steam heating system, and there's a lot of crud in these systems because the pipes and radiators are constantly corroding. When enough crud settles, the condensate has a tough time making it back to the boiler, so it backs up into the steam pipes, usually at the ends of the runs, and that causes water hammer.

I'll tell you about a piping arrangement called the

Hartford Loop a little later on. It's right there at the boiler, and if it's not piped properly, you'll get water hammer, usually near the end of the heating cycle. If this is the cause of your problem, you'll definitely need a knowledgeable pro to fix what the knucklehead did.

What's most important right now, though, is that you believe me when I tell you that water hammer is *not* normal. You may have had contractors tell you that it is ("Hey, it's steam. Wadda ya expect?"), but these are contractors that haven't done their homework. They shouldn't be giving you advice. Show them the door.

Getting rid of water hammer is sometimes easy, and sometimes not so easy. Once you figure out the cause, you look at your options. But the cause is always going to be mechanical. Remember that. There's no magic involved here, no need for an exorcism either. It's mechanical, and there's always a solution. *Always.*

Another sound that might come from your steam boiler is **rumbling**, and here I'm not talking about the sound that the burner might make (especially if it's an oil burner). I'm talking about a rumbling that can be so ferocious that it might even cause the boiler to start bouncing on the floor like Michael Flatley in *Riverdance*, which is never a pretty sight.

What causes the rumbling? Usually it's one of two things. The burner (that's the part that makes the fire) might be set up improperly inside the boiler and this is causing the water to boil more violently in one section of the boiler than it is in another. The too-rapid formation of steam can shove the water in the boiler this way and that and create so much hydraulic violence in there that the boiler begins to step dance. Oil-fired boilers are especially susceptible to this, but *only* if the person setting up the burner doesn't know his or her business – like a homeowner trying to save a buck (no offense meant). Oil burners have nozzles that are relatively easy to switch, and if the wrong nozzle gets into your burner bad things will happen. There's also the potential here for what oilheat people call a "puff back," which is a gentle-sounding word for a pretty horrible event. Say the word "puff back" to your insurance agent and watch him or her cringe. If you're not qualified to work on oil burners, please don't.

The other cause of rumbling is liming of the boiler. This happens after a steam system develops a leak and someone (or some automatic feed device) adds too much fresh water. Fresh water contains calcium and magnesium, two elements that have the strange property of coming *out* of solution as the water gets hotter. Calcium and magnesium attach to the inside of the boiler on the hottest surfaces and form a rock-hard material that most pros call **lime**. Get enough of this stuff in your boiler and pretty soon the water won't circulate properly. Pockets of steam will form and not be able to rise to the surface of the water because of the lime obstructions. The pulsing action of the steam as it forms and rapidly condenses beneath the surface of the water can be violent enough to cause your boiler to bounce. And rumble, of course. It's awe inspiring. Think Michael Flatley and get thee to a professional.

We'll talk more about noises as we go along, but what we should do now is try to figure out exactly what type of steam heating system it is that you have in your home. There are two basic types: one-pipe, and two-pipe. Which is yours?

Got One-Pipe Steam?

One-pipe steam takes its name from the way the radiators are connected to the supply pipes. Take a look at the drawing and notice that there's just one pipe connecting the radiator to the main steam line. The steam and the condensate have to share that single pipe. The steam goes up while the condensate flows down.

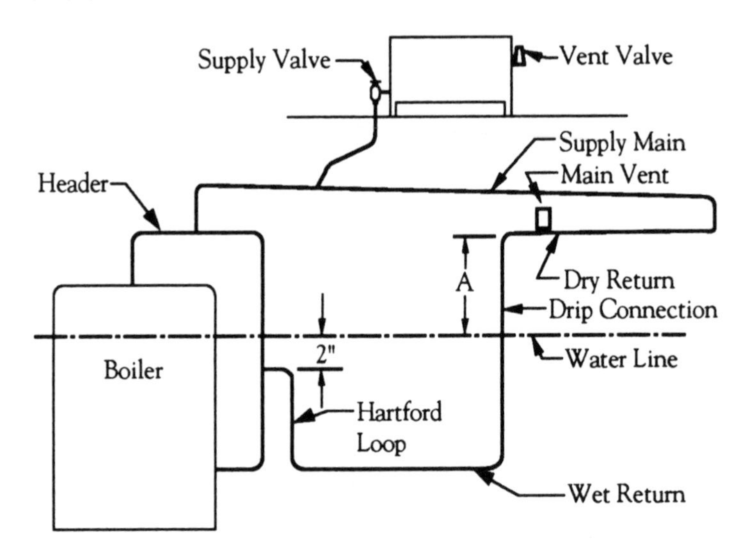

One-pipe steam systems are open to the atmosphere, as are all steam heating systems. They start out filled with air everywhere above the boiler waterline, and they end up filled with air when the system shuts off. As the steam forms and condenses, it pushes and pulls on the air like a big bellows. The air moves in and out of the system through the air vents, which means your system will corrode over time. The pipes are thick and so are the radiators and boiler sections so this will go on for many years, but as the system corrodes, the goop flows downhill.

Which brings us to the **wet return**. That's what we call any pipe that's below the boiler's waterline. These pipes are always filled with water. The steam can't go there because it would condense.

The opposite of a wet return is a **dry return**. You can see in the drawing that these are the pipes that are above the boiler's waterline. A dry return contains air, and then steam, and then air again as the system cools during the off cycle. There will also be condensate in the dry return, of course. That's the road that the condensate uses to get back into the wet return as it works its way back to the boiler to become steam once again. In a one-pipe steam system, the dry return is also the steam main. Condensate and steam share the same pipe, and that's another reason why we call it one-pipe steam.

The space marked "A" in the drawing is the vertical distance between the boiler's waterline and the bottom of the end of the dry return. Most pros call this space, **Dimension A**. It's a very important detail in a one-pipe steam system because it helps to put the condensate back into the boiler by giving the returning condensate a place to stack up and build static pressure (caused by the weight of the water). Dimension A has to be at least 28 vertical inches, and the Dead Men took great care to set this up properly. This is why, if your house is large, your boiler might be installed in a pit down there in your basement. The installer dug that pit so that he could get the proper vertical distance between the boiler's waterline and the bottom of the end of the dry return. Keep this in mind if you're replacing your old boiler. It can't just come up out of that pit. If you're having your boiler replaced, make sure the contractor checks Dimension A very carefully. A knowledgeable pro will do it automatically.

If you have one-pipe steam, there will be an **air vent** on each of your radiators. Here's what a typical radiator air vent looks like on the inside.

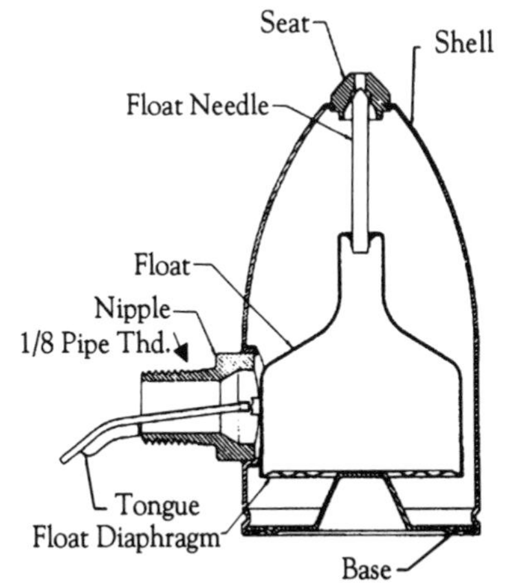

That float will pop up like a cork if water should surge into the vent from the radiator. That helps keep your walls clean. The float will also respond to heat because it's partially filled with a mixture of alcohol and water and sealed at the factory. Alcohol boils at a lower temperature than water. The manufacturer heats the float before placing it inside the air vent and that causes the alcohol to boil and turn to vapor. While the alcohol is boiling, the manufacturer solders the float closed and then allows it to cool. When the alcohol/water mixture condenses, it forms a vacuum inside the sealed float, which causes the flexible bottom of the float to bend inward, toward the center of the float.

The air vent manufacturer puts the float into the vent casing and sets the proper distance between the pin that sits atop the float and the hole in the top of the vent. Steam then pushes the air through the radiator and out the vent. When the steam reaches the vent, its heat causes the alcohol/water mixture to boil and the vapor that's produced increases the pressure inside the sealed float. That makes the bottom of the float pop out, driving the pin into the vent's hole and stopping the steam from leaking out. If crud from the system works its way into the space between the pin and the hole, the vent won't close and steam and water will escape.

If the steam pressure is too high, it can hold up the float and keep the pin stuck in the vent hole. That can keep the radiator from heating properly, and it's the main reason why the pressuretrol on your boiler has a cut-in and cut-out setting. The fluctuation in system pressure gives the float inside the vent a chance to drop so that more air can escape from your radiator.

Now, take a look at that small piece of metal called the **tongue**. It's right there at the vent's threaded inlet. The tongue helps water drain from the air vent and back into the radiator. It's a simple device, but an important one, sort of like the butter knife you might stick into a difficult bottle of ketchup to get it going. If the tongue gets bent, the vent will probably squirt and you'll need to replace that vent.

Some radiator air vents have adjustable air-release holes. These help to balance the system since big radiators contain more air than small radiators, and our goal is to get all the radiators hot at the same time on the coldest day of the year. You would use the fastest venting setting (usually the highest number on the adjustment dial) on the bigger radiators and the slower venting setting on the smaller radiators.

Some manufacturers offer a line of air vents that have fixed vent ports, but each vent in the series is faster than the previous one. Here again, the goal is to balance the overall system by balancing the release of air from the radiators. Big radiators need to vent air faster than small radiators.

And the radiator's location in your house has little to do with the vent you choose. If your system has main vents, the steam will favor the large pipes over the individual radiators when it first leaves the boiler because that will be its path of least resistance. The main vents allow you to very quickly fill all the pipes with steam so that the steam arrives at the inlet to each radiator at about the same time. From there, the radiator vents take over, venting the big radiators quickly and the small radiators more slowly. That way, everything gets warm at the same time.

Someone may tell you that you need to vent radiators that are close to the boiler quickly, and radiators far from the boiler slowly to balance the system, but this assumes that the system has no main vents. With main vents in place, the key to balancing a one-pipe steam system lies in matching the radiator vents to the radiators, regardless of where those radiators are in the house. Simply put, size matters.

If your radiator air vents are spitting, they're probably clogged with dirt and scale. You can try to clean them out by removing them (make sure the steam is off when you do this) and boiling them in a pot of vinegar. Vinegar is a mild acid that breaks down scale. It doesn't always work but it's worth a try before you go out to buy new air vents.

Now, there's an oddball air vent I want to mention

because you might have these on (actually inside) your radiators and not even know it. It's an internal air vent called In-Air-Rid and it was made by the American Radiator Company many years ago to satisfy folks who didn't like the look of air vents sticking out of the side of their radiators.

From the outside, all you see is a hexed plug in the upper part of the radiator's last section. If you look closely, though, (and if it hasn't been painted over), you'll see the words In-Air-Rid on that plug. The vent hole is the dot in the letter "i" in the word Air. One stroke of a paintbrush is all it takes to stop this vent from venting and your radiator from heating.

The cure is to use a straightened paperclip to poke the paint out of the hole. And if that doesn't work, get a pro to drill and tap the radiator for a standard air vent. You probably won't find a replacement In-Air-Rid vent. That company is long gone, but rest assured that as long as there are painters, manufacturers of standard radiator air vents will continue to prosper.

On the supply side of your one-pipe radiator is a **supply valve**.

Its job is to open and close access to the radiator. That's it. You can't have this valve halfway open because the steam and the condensate have to pass each other on their way in and out of the radiator. If you throttle the supply valve there won't be enough space for the steam and hot water to coexist, and the steam will win every time. It will just jam the water backwards into the radiator and cause it to come squirting out of your air vents. This will often be accompanied by some very memorable water hammer.

So keep the valve either opened or closed. And make sure the radiator is pitched toward the supply valve so that the condensate can flow from the radiator by gravity.

If the valve is leaking from the small nut that is on the valve's stem you may be able to repack this. Most hardware stores carry graphite packing. Loosen the nut when the system is off and wind some new packing around the threads. Tighten the nut. If this doesn't work, you can have the valve replaced. The new valve may not look exactly the same as the old valve, but they're still making steam supply valves so don't worry. A word of caution, though. Replacing a supply valve can sometimes lead to broken pipes so this is something you really should leave to the pros.

A one-pipe steam system will often add humidity to your home – whether or not you want it to do so. It all depends on how much steam escapes from the radiator air vents, which is a good reason to keep those air vents in working order.

Some folks like the extra humidity, though. I remember one homeowner who removed the air vent from the radiator in this room where he kept a lot of potted plants. The plants loved the warmth and humidity, but the boiler didn't care much for all the fresh water that entered to replace the steam that was lost to the room. Which is why I was there. His boiler had practically dissolved and he couldn't figure out why.

I also remember seeing a segment of *Eyewitness News* here in New York where a child died in a tenement building because Mom had removed the air vent from the radiator in his room, closed the door tightly, and then went out for the evening, leaving the kid alone. What a tragic way to learn how steam has the power to displace air.

If you want to add humidity to your home and you have steam heat, please don't use the steam from the radiators. For years, people have put trays of water on top of their steam radiators, or hung galvanized steel receptacles behind the radiators. The heat from the radiator causes the water to evaporate and that does a nice job of raising the indoor humidity level on those frigid days Cold air that enters your home whenever you open a door can't hold as much moisture as warm air and that's why you'll notice more static electricity in your home on very cold days, and why your lips or nose may chap. The indoor humidity level just isn't what it should be. The Dead Men knew what they were doing when they placed those trays of water on their radiators. Feel free to do the same.

If you have one-pipe steam heat and you're hypersensitive to odors, watch what you allow any serviceperson to put into your boiler. Chemicals that are designed to slow corrosion or help the quality of the steam sometimes come with an odor that can be very noticeable when the one-pipe steam air vents are doing their work. The smell from the boiler water usually winds up in the room with you.

A common additive to a steam boiler that's acting up is vinegar. Vinegar, is a mild acid that can lower the boiler water's pH. That helps to get rid of what we call **foaming**. I'll tell you more about foaming later on, but for now, just know that if someone adds vinegar to your boiler it can lead to a temporary vinegary smell in the rooms upstairs, and if that's going to bother you, don't let them do it. Ask the serviceperson to use another type of mild acid. Orange or lemon juice, for instance.

And speaking of odors, a true story for you. A tenant in a fancy co-op apartment building calls to say that there's a terrible odor coming from the radiators in her apartment. Other tenants are smelling the same odor. The consultant arrives with a fresh nose and sure enough the place stinks. And it's a very distinctive odor, no mistaking it at all. The place smells like the men's room in a seedy bar. No mistaking this odor. Nope.

A bit of poking around and some friendly questioning leads to the building superintendent. He lives in the basement and has no radiators. He gets his heat from the steam pipes overhead. He's not happy with the gratuities received during the recent holiday season. So he collects his urine each day in a big Maxwell House coffee can, removes the boiler's relief valve each night, pours the urine into the hole and replaces the relief valve. Then he starts the burner. The building has one-pipe steam and the odor just wafts upstairs into the fancy apartments.

So there.

You can't make this stuff up.

All in all, one-pipe is the simplest of steam systems. The steam goes up, the water comes down. Here are some things to watch out for, though.

✓ Dirty or oily water

✓ Pipes that aren't properly pitched or insulated

✓ Air vents that are clogged or not the right size

✓ A wet return that's clogged with system goop

✓ Radiators that pitch the wrong way

✓ A replacement boiler that's not piped correctly

✓ Disgruntled building superintendents

Those are the main causes of one-pipe steam heating woes. Address these and you can peacefully coexist with your system. If you get stuck, stop by www.HeatingHelp.com and post your question on The Wall. That's our very active bulletin board. We've got some of the best steam heating people in the world hanging out there. Stop by; we can help.

Now let's take a look at that other type of residential steam system.

Got Two-Pipe Steam?

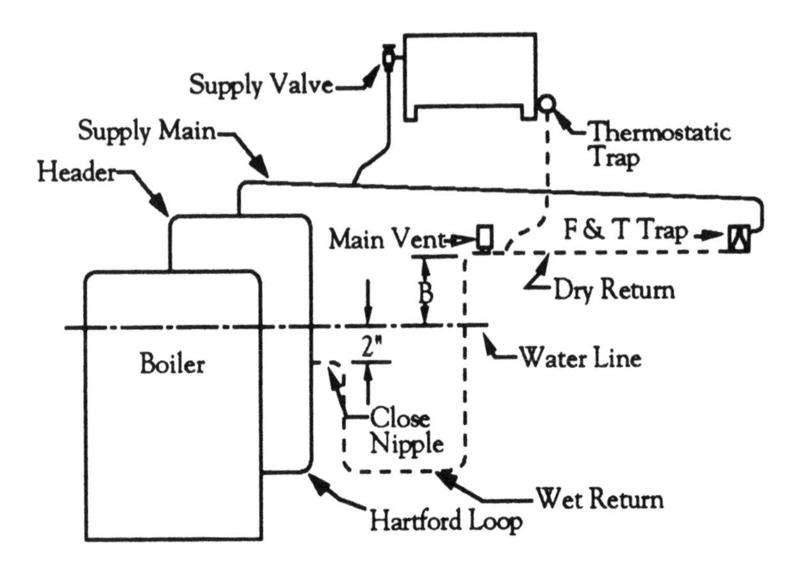

Two-pipe steam, like one-pipe steam, takes its name from the number of connections at each radiator. We have a supply pipe that handles only steam, and a return pipe that handles only air and condensate. Two-pipe radiators don't have any air vents on them and that was a big selling feature in the early days of steam heating when air vents weren't that reliable. Folks who had two-pipe steam didn't have to worry about steam or water squirting out of a radiator vent and onto their walls and curtains. And there was also less odor with these systems since the air moves through the radiators and into the return lines and leaves the system through one or more large main vents down in the basement.

Two-pipe steam, as the Dead Men installed it in homes, also went by the name of **Vapor Heating**. In a Vapor system, the required system pressure is never more than eight ounces per square inch (hard to believe, eh?). There was also **Vapor/Vacuum Heating**. That was the name they gave the system if it operated between eight ounces of pressure on the high side, and down into a vacuum on the low side. I'll explain more about how that works in just a little while, but first, let me tell you about some of the features of two-pipe steam.

As with a one-pipe system, we begin with water in the boiler, and in all the piping that's below the boiler waterline (down there in the wet returns). The pipes above the boiler's waterline, as well as all the radiators, contain air, and we have to get rid of that air before the steam can get upstairs to make things cozy. Remember, there's never been a steam system that didn't need to be vented.

Oh, and just so you know, there's no difference in the boiler when it comes to these systems. One-pipe and two-pipe steam systems use the same type of boiler. The difference between the systems lies in the piping and the radiators.

So the burner comes on and boils the water and steam begins to leave on its way to your radiators. As with one-pipe, the steam pushes the air ahead of itself. The air goes through the main and up the supply line to each radiator. It moves past the supply valve, through the radiator, and then enters the return line by passing through the steam trap (we'll get to those in a second). Once in the return, the air moves toward the main vents and leaves the system. With most two-pipe systems, the air will reenter through these same main vents when the boiler shuts off and the steam condenses. Nature hates a vacuum and the condensing steam just sucks the air right back in. That's not a bad thing.

So, as you can see, two-pipe steam is also an open system, just like one-pipe steam. Air is constantly moving in and out, which causes the metal that makes up the system to rust over time. All steam heating systems do this. They're constantly corroding and always in need of love and attention.

Now, take a look at that riser that goes up to the radiator supply valve. Its job is to deliver steam. That's it. There won't be any condensate rolling backwards down that riser as there was with our one-pipe system. That's another advantage of two-pipe steam. Because there's less counterflow, there's less chance for water hammer. The rich folks who owned those early systems must have liked that a lot.

Part of the appeal of a two-pipe radiator is that steam enters from the top and travels downward, pushing air ahead of itself and out the steam trap. As the steam gives up its latent heat energy and condenses, the water rolls down the inside of the radiator and warms it even more. Because of this, two-pipe radiators often give off a more-even heat than one-pipe radiators. And since the steam and condensate don't have to pass each other in the supply valve, you can throttle the supply valve on a two-pipe radiator, limiting the amount of steam that can enter the radiator on those milder days.

The outlet side of the radiator has a thermostatic steam trap. Its job is to first allow air to pass, then to close when steam arrives, and finally, to reopen when condensate forms, allowing the condensate access to the return line so that it can work its way back to the boiler by gravity. Here's what the inside of a thermostatic radiator trap looks like.

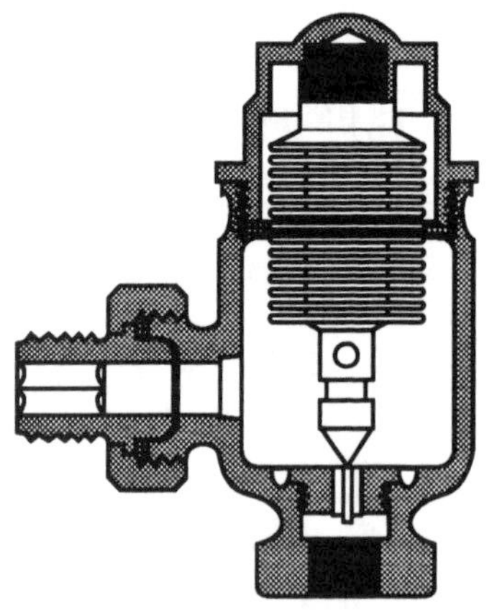

The inlet is on the side and the outlet is on the bottom. The piece in the middle that looks like a bellows is called a **thermostatic element**. It contains a mixture of alcohol and water, and it's sealed under a vacuum, just like a can of ground coffee. The manufacturer performs this trick by partially filling the element with the alcohol/water mixture and then heating it until the mixture boils (at a temperature lower than 212° F). While the mixture is boiling, the manufacturer will solder closed the opening in the bellows and remove the heat. The alcohol/water mixture condenses in the now-sealed element, and that causes the element to collapse like a closed accordion. The manufacturer will then put the element inside the trap's metal body. The bellows is tucked-up and that means that the trap will normally be open. Only steam can close it.

Once on the job, air will pass by the open element, and so will water. But when steam arrives, it causes the alcohol/water mixture to boil inside the sealed element. The pressure created by the boiling fluid pushes the element's bellows downward, and the pin that's attached to the bottom of the element will move into the trap's seat, stopping the steam in its tracks. The steam will have to condense, and the condensate will have to cool the fluid inside the bellows by about 15 degrees Fahrenheit before the trap will reopen.

Radiator traps are important in two-pipe systems because they keep the steam from getting into places where it doesn't belong. If steam gets through a radiator and into the return line it can trap air in nearby radiators and cause them to remain cold. The return lines are also much too small to accommodate steam so you'll often wind up with water hammer, and water hammer can be very destructive. One bad radiator trap can cause the death of other traps by hammering them to a metallic pulp.

How long do thermostatic radiator traps last? Probably not as long as you think. Most trap manufacturers say that the average lifespan of a thermostatic trap's element is about 10 years, and that's if they're in a well-maintained system. Ten years. That's it. When was the last time you had your traps checked?

I thought so.

The problem is the element doesn't cry out when it dies. It just stops working and the funny part is that the radiator with the bad steam trap will probably continue to heat. It's the nearby radiators, the ones with the good traps, that will stop heating. Ironic, isn't it? The bad trap passes steam into the return and that sends you looking for the cause of the problem in an area where it ain't. Steam's funny that way. The problem and the cause of the problem will never be in the same place.

So how do you check these traps to see if they're working? Well, the simplest way is to take their temperature. To do this, you'll need a good thermometer that can measure the surface temperature of a pipe. Check the temperature at the inlet of the trap when the steam is up, and then check it a few inches away from the outlet of the trap. If the trap element is good you should see a 10-15°F. difference in temperature from one side of the trap to the other. If the trap isn't working, there will be hardly any temperature difference because the steam will be passing right through the trap and into the return.

Can you change the elements yourself? You can, but only if you're handy, strong, cautious, courageous and have the right tools. You'll have to be able to unscrew the cap on the top of the trap. That's where the tools come in. Some of the older traps have special wrenches and that's not something you'll find at your local home center. Without the special wrench you stand a good chance of snapping a pipe below the floor and that's never a pleasant experience.

As for the replacement parts, you should be able to find them at any good plumbing & heating supply house. Many of the trap manufacturers are no longer in business, but there are companies that make repair kits to fix most any trap (you'll need to know the size and model number of the trap when you get in touch with them). Here are some names of folks who supply universal repair kits for steam traps:

* Barnes and Jones 1-781-963-8000
* Hoffman Specialty 1-773-263-7700
* Tunstall Associates 1-800-423-5578

All good people. As for me, though, I'd rather let a pro take on the responsibility of dealing with what might be weak pipes around those old steam traps. If he breaks them, he can't blame me.

But if you're the type that likes to live on the edge, by all means, go for it. Just don't forget to turn off the steam before you open up anything. And don't do the work during the winter.

And remember I told you to call a pro.

Vapor devices

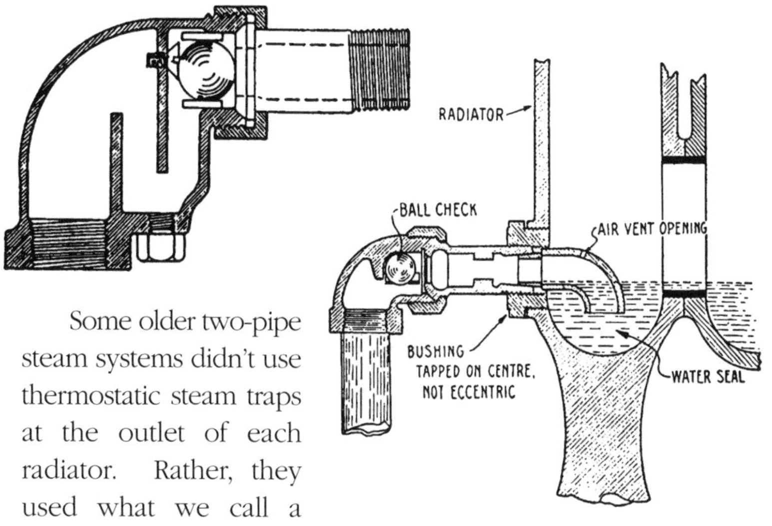

Some older two-pipe steam systems didn't use thermostatic steam traps at the outlet of each radiator. Rather, they used what we call a **return elbow trap** or **vapor orifice**. Vapor systems ran on extremely low pressure, never more than eight ounces. Steam at that low a pressure travels very fast and the vapor device at the radiator outlet would ensure that all of the steam that entered the radiator would condense and not be able to enter the return. The return elbow trap or orifice contains a very small hole through which the condensate has to squeeze. As the condensate backs up at the orifice, it blocks any steam vapor that tries to get into the return. Since the steam is at very low pressure, it doesn't have enough power to shove the condensate out of its way. It's a remarkably simple device, but if dirt gets into the orifice (which is very likely) then the radiator won't heat well. Again, you have to be able to open the return elbow trap to get at the orifice to clean it, and this may or may not result in an Adventure in Steam Heating for you, depending on the age of the pipes, and your own age, of course.

I'd call a pro.

My book, *The Lost Art of Steam Heating* (available at HeatingHelp.com) has an extensive section on these old vapor heating devices. Most go beyond the range of what I think a homeowner should be messing with, but I'll refer you to that book (which I wrote for contractors) if you're interested in learning more.

The most important thing, though, is that you don't remove, or allow anyone to remove, any vapor-system component unless you're absolutely certain of its function in the system. Every part is important and removing one might cause the whole system to shut down.

End-of-main steam traps

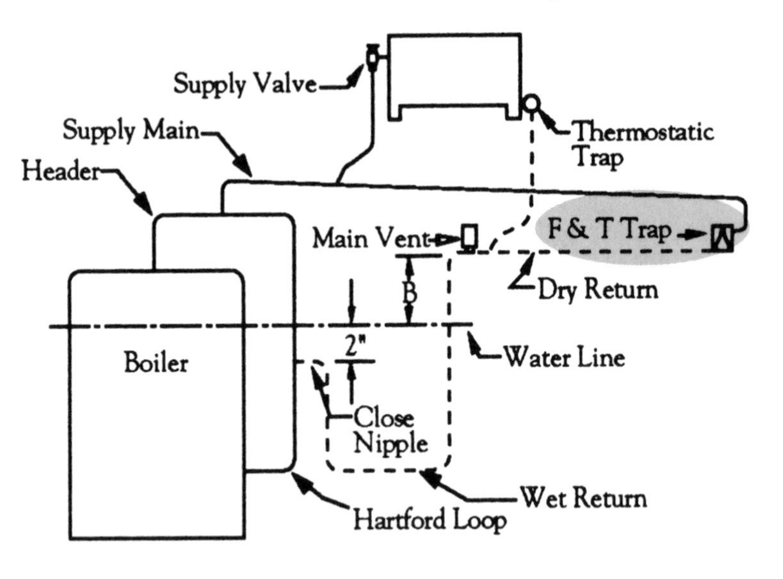

Here's that overview of the two-pipe system again. I've highlighted the trap near the end of the steam main. Its job is to keep the steam that's in the main from crossing over into the horizontal return pipes that are above the boiler's waterline (again, we call those the dry returns). The return lines contain the condensate that's draining from your radiators. They also contain the air that the steam is pushing through the radiators. Near the ends of the dry return lines there are main air vents and if the steam were allowed into these lines it would close those air vents and your home would stop heating. You'd also have a lot of water hammer because those return lines contain condensate and they're simply too small to accommodate both steam and condensate.

The trap that gets assigned to this key position in the system is called a **Float & Thermostatic trap**. More often, we just shorten the name to the simpler, **F&T trap**. Here's what one looks like on the inside.

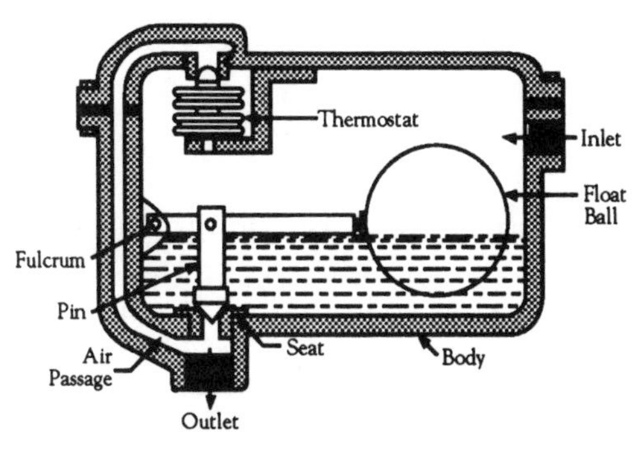

The float is attached to a lever and a pin that seats itself in the trap's outlet. That float is normally down and the trap is always closed unless condensate accumulates inside of it. The condensate causes the float to rise, lifting the pin from the seat so that the condensate can drain. When the steam enters the trap, the float remains in the down position and the steam has no choice but to condense and turn into water, which will lift the float and open the trap.

The thermostat part of an F&T is very similar to what's inside a thermostatic radiator trap, but the difference here is that the thermostat in an F&T doesn't have to deal with condensate. All it has to worry about is the air that the steam pushes ahead of itself. The thermostatic element, like the one in a radiator trap, is normally open. Air passes by, but when steam arrives, the alcohol/water mixture inside the thermostatic element boils and the vapor pressure created inside the sealed element causes the bellows to expand like an accordion and that keeps steam from entering the return line.

Checking F&T traps is definitely a job for a pro because F&Ts don't work on temperature difference. The float will open and dump condensate that's at steam temperature.

All that's removed is the latent heat, and a thermometer can't sense latent heat. The temperature at the outlet side of the trap will be the same as the temperature at the inlet side of the trap.

A pro will open a line just past the trap while the steam is up (a dangerous task). He'll examine what's coming out of the trap and determine from this whether the trap is working or not. If the end-of-main F&T traps in your two-pipe system have never been checked, and if your system is noisy or slow to heat, my guess is that those traps need to be repaired or replaced. It's a good investment, and the results will amaze you.

(A little food for thought here)

I've had people tell me that they haven't fixed their steam traps (both radiator traps and F&T traps) because they can't afford it. I always tell those people that they can afford it. In fact, they're already affording it, but they're making the payment to their fuel supplier rather than to their contractor. I've seen enough steam systems to know that this is true. The thing is, you never notice that your fuel bill is higher than it should be because it's been high for such a long time, and it didn't happen overnight. You're making a payment toward the repair of those steam traps every month by way of abnormally high fuel bills – but the traps never getting fixed.

That's like buying a new car, getting the payment book, and then never picking up the car. Crazy, right?

Can't afford to get those traps fixed? Sure you can! You're already making the payments. All you have to do now is shift those never-ending payments to a knowledgeable contractor. He only needs to be paid once.

Return pumps

If your home is of the Extra-Large variety, your two-pipe steam system (and sometimes even your one-pipe steam system) might have a **condensate return pump** or a **boiler-feed pump** sitting on the floor near your boiler. Most homes don't need one of these devices but if you do have a pump it's there for one reason only – there's not enough vertical distance between your boiler's waterline and the bottom of the lowest steam trap. Here, take another look at a two-pipe system without the return pump.

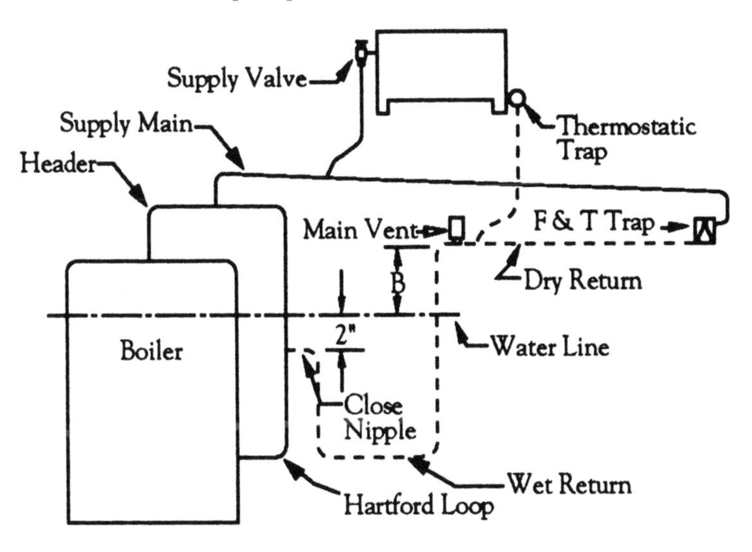

Take a look at the part of the system that I've marked **Dimension B**. This is the vertical space between the boiler's waterline and the bottom of the lowest steam trap, wherever that may be (it's usually the one at the end of the main). Since your system has those steam traps on the radiators and at the ends of the steam mains, there's no steam pressure available to help get the returning condensate get back into the boiler. Steam traps trap steam, so two-pipe systems can't work the same way that one-pipe systems do. You'll recall that with the one-pipe system, we had **Dimension A** (the vertical distance between the boiler's waterline and the bottom of the lowest steam-carrying pipe). Dimension A contained "leftover" steam at the end of the run, and that steam's pressure combined with the weight of the water as it stacked in Dimension A to put the water back into the boiler. Since two-pipe systems don't have any "leftover" steam pressure available to the returns (because of the traps), all we have going for us is the static weight of the water in Dimension B. And you need 30 inches of vertical space between the boiler waterline and the lowest steam trap for each pound of steam pressure inside the boiler to make things work by gravity.

So if your two-pipe system is operating at, say, two pounds per square inch pressure, you'll have to have 60 inches (five feet!) of vertical space between your boiler's waterline and the bottom of the lowest steam trap.

Many basements don't have that much space available, and that's where the return pump comes it. It provides the pressure that gravity can't.

And it's a very simple device. Here, look.

What we have is a cast iron (or steel) receiver that serves as a collection point for the condensate that's draining from the system. The water flows into the receiver and lifts the float, which then trips an electrical switch and starts the pump. The pump moves the water from the receiver into the boiler, and as it does this, the float falls and shuts off the pump. Simple.

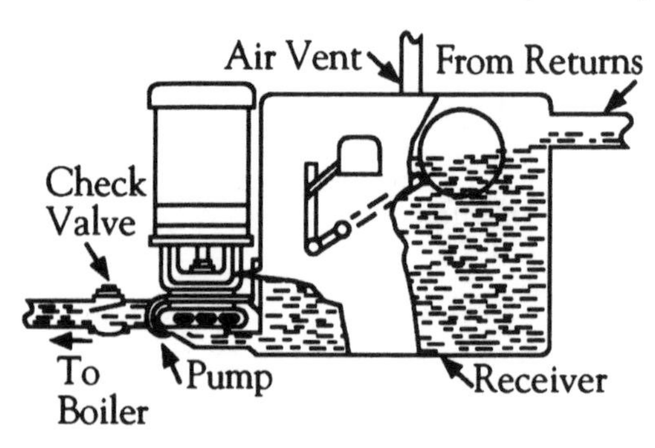

There's a check valve at the discharge side of the pump and its job is to allow the water to flow just one way (into the boiler). A check valve is a special valve that has an internal flapper. Water can only go one way through a check valve. If it tries to back up, the flapper closes and stops it. It's like a turnstile. Water can go only in one direction. Without the check valve, the water in the boiler would simply flow back into the receiver every time the pump shut off. And that would turn the pump back on. And so on. Before you know it, you'd be replacing that pump.

There should also be a throttling valve just after the check valve. A pro will use this to set the pressure that the pump uses when it puts the water back into the boiler. Without the throttling valve, the check valve might chatter like a castanet whenever the pump runs.

Now, the one bad thing about a condensate pump is that it stops and starts in response to that float-operated switch inside the receiver. That switch has no idea whether the boiler needs water. It just responds to the water level inside the receiver. If your system has an automatic water feeder (more on those coming up soon) there's a good chance that the condensate pump can flood your boiler. If that's happening, there's most likely something wrong out in the system. Something is slowing the return of condensate to the pump's receiver. It might be accumulated dirt in the return lines, or it could be failed thermostatic radiator traps or end-of-main traps that are pressuring the return and slowing the water as it tries to return to the pump's receiver by gravity. If this is the case, you may see steam coming from the condensate pump's vent line. That's the vertical pipe that extends up about five feet. It's open on the top. Boiler-feed pumps also have these open vent lines. It's how the air leaves the system when a system has a return pump. Steam should not be coming from that vent. If it does, someone has to track down the cause. *Never* put a cap on that vent line. If you do, the receiver will become pressurized and it can explode with the force of a bomb. It happens every year, and with disastrous results. Please don't put *anything* in that vent line.

The other type of condensate return pump is called a

boiler-feed pump. These have larger receivers than the ones you'll find on condensate pumps, and the control that starts and stops the pump is mounted on the boiler, not inside the receiver. Take a look.

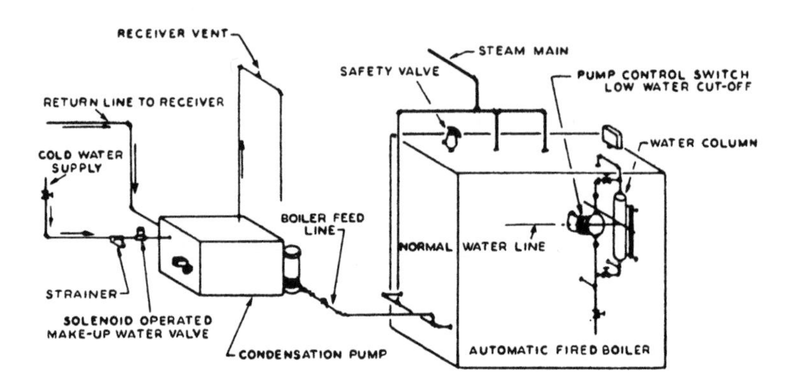

If you're replacing your old steam boiler, your contractor may talk to you about adding a boiler-feed pump to the system. He'll bring this up because modern steam boilers contain less water than the boilers they're replacing. There's a chance that the new boiler might run out of water before the condensate returns from your system. That's where the boiler-feed pump comes in. It holds a reserve of water in its receiver, and it has enough space to accommodate the returning condensate. And since the pump controller is mounted on the boiler rather than inside the receiver, the only way water can enter your boiler is through the boiler-feed pump. And that will happen only when the controller senses a need for water inside the boiler. You normally wouldn't have an automatic water feeder if you have a boiler-feed pump, so that eliminates a lot of hydraulic confusion as far as the feed water is concerned. All the feed water entering the system comes through that pump.

So your contractor may recommend a boiler-feed pump if you're replacing a big old boiler with a small new one. Whether or not you need the pump is another matter. Not all jobs do, and this would be a good time to seek a second opinion from another pro. Often, if the new boiler is piped well and cleaned well, if the radiator traps are working as they should, if the return lines are clear of muck, and if the mains are properly insulated, you won't need that boiler-feed pump. Unless your house is huge that is (and here I'm talking mansion).

I say this because there's more involved in this job than just the pump. The pro will also have to add steam traps to key areas, and some return lines may need to be replaced or rerouted. It's a relatively expensive proposition and if you don't need it there's really no reason to go for it. Some contractors will try to put one on every job, though, so be forewarned. It pays to get a second opinion.

Choosing a Boiler (or dealing with the one you already have)

I've had the opportunity to visit several iron foundries over the years and after each visit I've gone home and felt real good about being a writer and not a guy who pours molten metal for a living. The folks that work there, in what looks like Hell's lobby, perform brutal, primal work. It involves fire and elements from the center of the earth, and the heat coming off those huge caldrons can take your breath away.

The people who do this difficult work pour the molten iron between two molds that are made of a special, highly compressed sand. The molten iron chases around the mold and it's mesmerizing to watch the process. There's this sudden explosion of hot gases that occurs just after the iron enters the mold and it's startling the first time you hear it. I jumped a foot! It's part art, part science and well worth watching if you ever have the opportunity. You will go home and kiss your own job on the mouth, no matter what you do for a living.

When the iron cools (and that takes hours), they put the iron castings into a mechanical shaker that knocks the sand out of the inside and away from the outside of the casting. What they wind up with is an individual boiler section, made of cast iron and hollow on the inside. That's where the water will be when it's doing its job in your basement.

There are hundreds of these short, thimble-like iron pins cast into the outside surface of each boiler section. When they put two sections together, the heads of the iron thimbles will almost touch each other. The heat from the oil- or gas-fired burner will flow past these pins and transfer through the iron and into the water, causing it to boil. The boilermaker will machine each section so that it marries to the next, and the next, like slices of bread in a loaf. Once assembled, this will be what we call a cast iron sectional boiler (for obvious reasons). It's the most common type of steam heating boiler, and probably what you have in your home.

Courtesy of Weil-McLain

They attach the burner near the bottom of the assembled boiler. The burner might fire into a chamber inside the boiler, or burn from metal rods that are under the boiler (like in a gas barbeque). The hot gases flow upward between the sections and enter the chimney where they continue to rise by convection up into the air.

The connected boiler sections have an area above the boiler's waterline (you can see the waterline in the gauge glass) that we call the **steam chest**. This is the place where steam breaks free of the surface of the boiling water. At the top of the steam chest there will be one or more large openings through which the steam will leave the boiler. Into these openings, the installer will screw pipes called **risers**. These risers connect into a horizontal pipe called a **header**. The header directs the steam to the **steam mains** that will carry it out into the system and toward your radiators.

A steam boiler is a lot like a pot of water on a stove, only more complicated. And the proper name for this item is a **boiler**. Please don't call it a furnace. A **furnace** is a unit that heats air, not water. Contractors chuckle at homeowners who call boilers "furnaces." Use the proper terms when you're talking to contractors and they'll treat you with more respect. Professional contractors appreciate educated consumers because an educated consumer makes a pro's job easier. That consumer knows what's involved in getting the job done right. You're well on your way to becoming one of those educated consumers.

Okay, a bit about **burners**. These can fire natural gas, propane or oil. The pro will size the burner to the **connected load** on the steam boiler. That load includes all the radiators and all the pipes in the system. He'll carefully measure the connected load before giving you a price on a replacement boiler. If he doesn't take the time to do this, you really shouldn't do business with him.

One thing I'm *not* going to do in this book is tell you how to setup or clean burners. I'm not going to do this because you're *my* customer and I'd like to keep you alive for awhile. Modern burners need to be set using instruments that most homeowners don't have. The folks who use these instruments also attend many hours of training before they head out into the field, and they keep taking courses as the years go by because equipment constantly changes. They do this to protect your life. If not set properly, a burner will not only waste fuel, it can create deadly carbon monoxide, a gas that has no odor, color or taste. *Any* sort of burner – natural gas, propane, or oil – can produce carbon monoxide. Please let a skilled professional serve you when it comes to your burner. It's money *very* well spent. Thanks.

Now, a quick word about system piping and how it relates to the fuel used. In the old days, most steam boilers ran on coal, and the coal fire was continuous. When these systems were converted to burn oil or gas the flame became intermittent because you can start and stop an oil- or gas-fired burner with a thermostat. This often changed the way the steam system performed. Some rooms would be too hot while other rooms stayed cool. It all depended on how the piping ran around the basement. If a single main ran around the entire basement (and this was typical piping for coal) the pro who later converted the fuel from coal to gas or oil would usually add additional vents along the main to speed the steam on its way around the basement.

Years later, when gas and oil replaced coal as primary fuels, the installers would use more than one main around a basement. One main would go to the left while another went to the right. Depending on the shape of the house, these mains may be of different sizes, the goal being to balance the pressure drop so that all the mains would heat evenly (another reason why you can't arbitrarily change pipe sizes if you decide to move things around). The ends of these mains would typically drop below the boiler's waterline at the opposite side of the basement and return across (or under) the basement floor through a single pipe.

See? Piping and burners have a relationship when it comes to steam. As does just about everything else.

How manufacturers rate boilers

If you look at a boiler manufacturer's catalog you'll notice that they use several types of ratings for each boiler. There's the **Input** rating, the **Gross Output** rating, which, to further confuse things, some refer to as **D.O.E**. (Department of Energy) **Heating Capacity**, and then there's the **Net Output** rating. To make things even more confusing, some of the ratings are shown as Btuh (British thermal units per hour), while others are listed as Square Feet EDR (Equivalent Direct

Radiation). And there might be a column that shows the boiler's ability to burn fuel oil (listed as Gallons Per Hour), or natural gas (shown in Therms).

So what the heck's going on here?

Here's the deal. If a contractor shows you the boiler manufacturer's product literature you're going to see three basic columns. First there's Input. This column has to do only with the fire in the belly of the boiler so this is where you'll find the fuel-burning ratings in Gallons Per Hour (for fuel oil) or in Therms (for gas). What you're seeing in this column is the amount of heat that the fire is putting into the boiler. You PUT the fire IN the boiler and that's why they call it Input. Marvelously simple, isn't it? Sure, it is. You can figure out most of this stuff just by turning around the names of things. Air vents, well, they vent air! And steam traps trap steam. Condensate pumps pump condensate (surprised?). Input puts in. Output puts out. See? Simple!

Convinced yet?

Yes, I thought you would be (well, almost).

And in case you're wondering, burning one gallon per hour of Number 2 fuel oil (that's what residential oil burners use) yields 140,000 Btuh. One Therm of natural gas burned produces 100,000 Btuh. If you're using propane, that's equal to 92,000 Btuh per gallon.

Anyway, not all the heat that goes into the boiler from the fire winds up in the water. Some of that heat goes up the chimney and some of it goes out through the sides of the boiler and into the boiler room. We call what's left after these loses the Gross Output. Now, this term can be a bit confusing because Gross usually implies that you're dealing with the whole enchilada, as in Gross Income (which means **before** taxes). But in the World of Steam Heating, Gross actually means, "what's left over" instead of "what you start with." Or to put it another way, Gross means **after** taxes. Taxes, in this case, being the price you pay when you send some heat up the chimney and through the boiler's jacket. The important thing to remember is that Gross Output is the amount of heat that rides on the steam that's flowing out of your boiler. It's the heat that's actually available to the system (and maybe *that's* why we call it Gross).

The difference between the Input and the Gross Output represents the combustion efficiency of the boiler. For instance, if a boiler has an Input of 200,000 BTUH and a Gross Output of 160,000 BTUH, that boiler would be running at 80% combustion efficiency (since 160,000 is 80% of 200,000). It's not hard to figure this out. Just divide the big number into the small number and then multiply the result by 100 to get a percentage. Generally speaking, the higher the combustion efficiency, the noisier the salesperson.

And this brings us to **Net Output**. In the real world, Net is what you're left with after taxes. In the World of Steam Heating, Net is always going to be less than the Gross Output because there are two things going on out there in your system. First, we have the heat loss caused by the piping. It's going to take a certain amount of heat to bring all those pipes from room temperature up to the temperature of the steam that's flowing through the pipes. And this is where things can get a bit more confusing. Are those pipes within the living space? If they are, the heat really isn't lost. Are those pipes insulated? Insulation reduces heat loss. How well are the pipes insulated? How large are they? All these things make a difference.

And this piping loss has nothing to do with the heat loss of your home. It has to do with raising the temperature of the hundreds (or perhaps thousands) of pounds of steel (the weight of the pipes) from room temperature to steam temperature so that the heat can arrive at your radiators in the form of steam. The boiler's ability to produce steam has to match the system's ability to condense steam. It's as simple as that.

Something else to consider. As steam heating systems evolved, the average amount of piping in a typical home changed. One-pipe systems contained less metal than two-pipe systems. This affected the **piping-pickup factor** that boiler manufacturers used. Piping-pickup is another of those technical terms. It refers to the amount of extra capacity a manufacturer has to put into a boiler to "pickup" the load of the piping when the system first starts. The larger the pipe, the bigger the pickup factor. For instance, in 1940 the factor that they used to get from a boiler's Net rating to its Gross Output was 1.56. In other words, they figured out the building's heat loss. Then they selected radiators that could overcome that heat loss. They then took that installed radiation load and multiplied it by a factor of 1.56 to get to the boiler's Gross Output rating. That's a pretty hefty increase, and it was because of the large pipe used in the old days, and also because the manufacturers were being *very* conservative. They didn't want to have unhappy (read, cold) customers.

By 1945, however, the steam heating people had some serious competition from the people who were selling hot water and warm air furnaces. The steam boiler manufacturers realized that they were probably being too conservative with that 1.56 piping pickup factor, which was putting them in a non-competitive situation, so they reduced the Pickup Factor to 1.33. And that's where it remains to this day. It's the standard nowadays, and in most cases, it's perfectly fine.

So, if properly sized, your boiler has the ability to heat all the radiators in your home, plus a load equal to one-third of the total radiation, and that one-third is there to bring all the pipes up to steam temperature.

And this is another good reason to make sure your steam pipes are well insulated. When you remove insulation, the heat loss of the pipes will be five times greater. That can leave you with an effectively undersized boiler, and cold rooms. All of these boiler ratings are based on *insulated* pipes. If you remove asbestos and don't replace it with fiberglass, you'll be buying more fuel than you actually need to heat your home. Pipe insulation will *always* be cheaper than fuel. You only have to buy the insulation once.

The thing about those older boilers

When steam heating was new, back at the turn of the 19th Century, coal was the fuel of choice. It burned hot and it burned long. A good coal-fired boiler would stay lit for eight to 10 hours on a load of coal. Those early boilers contained cast iron grates that were similar to what you have in your barbeque, but much thicker and far sturdier. The person tending the fire (and it was almost always the woman of the house) had to know how to start the fire and spread the coals so that the air could enter from below and keep the home fires burning.

And I know that it was the women making the fires because my antique engineering books tell me so. The writers of those books made a big deal about this gender thing, and most suggested that a smart heating engineer who is choosing a boiler for a house will add up to 75 percent more capacity to the coal-burning grate area than is actually required to get the job done. Can you guess why?

It was because the man of the house might try his hand at building a fire on the weekends.

The writers of the old engineering books flat out wrote that the men didn't know what the heck they were doing when it came to making a proper coal fire (it was women's work). The writers figured that if they gave the husbands more room in which to play, there would be less chance of a lousy fire on the weekends, and fewer complaints to the heating engineer.

Seventy-five percent extra space in the combustion chamber for the husband. Because he didn't know what he was doing.

How about that?

And there's something else you may not know, something that had a huge impact on the development of house heating back in the early days of the 20th Century. It was the Spanish Influenza pandemic of 1918/19. The flu killed 40 million people worldwide.

Roll that number around in your mind for a moment.

Forty *million* people.

This was the worst disaster in human history but you probably never learned about it in school. It was an event so horrific that it was literally erased from the common memory. It killed so many people during that winter that the editors of *The Ladies Home Journal* magazine coined a new term during the summer of 1919. They decided to rename the parlor the "living room" in honor of those who survived. "Parlor" had taken on such a sad connotation because that was where most families laid out their dead.

The Spanish Influenza disappeared as suddenly as it had arrived but it affected people for years afterward. During the summer of 1919 no one was allowed out on the street unless they were wearing a surgical mask. If you go on the Internet you'll find photographs of Major League baseball games where every player is wearing a mask, and so is every fan in the stands.

This was the scariest virus of all time, the perfect virus, and it was airborne. People became deathly afraid of the air in their homes, in their workplaces, in their houses of worship, everywhere. They began to open their windows to let in fresh air and this had a dramatic effect on central heating. You can see it in the engineering books published after 1920. The authors wrote of the Fresh Air Movement and cautioned engineers to specify boilers and radiators that will be large enough to heat the building on the coldest day of the year – with the windows open.

If a contractor bases the size of your new boiler on the label that's on the old boiler, there's a very good chance he'll be sizing a boiler for you that's large enough to heat your home with the windows open. He'll also be accommodating a long-dead husband who didn't know how to make a coal fire on the weekends. These are vestiges of American Heating History. There's no reason why you should have to embrace them, though.

But wait, it gets even wackier. The Dead Men oversized those early boilers for good reasons, and they installed them well so that they would last for decades, which they did. What changed as time went by, however, was the fuel. Coal was never cheap, and once the oil burner arrived in the late 1920s, oil heat looked very attractive. The Spanish flu had vanished and people got on with their lives. Oil was cheaper than coal, and it was an automatic fuel. That made quite an impression on the woman who had been keeping the home fires burning. Imagine, a boiler that takes care of itself! And oil is a lot cleaner than coal. Switch to oil and there would be less housework. And you can reclaim the space in the basement where the coal bin used to be. Get rid of the coal dust and your kids would be healthier. Yes, oil was the way to go!

The challenge for the contractor, however, was how to size the new oil burner for the old coal-fired boiler. To make the switch in fuels, he would have to remove the coal grates from the boiler, fill the base of the boiler with sand, and then mount the new oil burner on the cleanout door of the old boiler. It was more art than science.

And to make it even tougher, there was no reliable conversion formula to go from coal to fuel oil because the heat value of the coal varied depending on the type of coal and where it came from. The oil dealer didn't want to pick a burner that would be too small to get the job done because then his customer would complain, so he figured in a generous safety factor.

You following this? We begin with a coal-fired steam boiler that can heat the house with the windows open. To this we add 75 percent because of the husband, and then the oil man adds a margin for safety to that.

And that's the way Americans heated for years. Too hot? Open the window. Oil was cheap and many of those boilers lasted through the 1960s and into the early-1970s. And then 1973 rolled around, and with it came the first OPEC oil embargo. The price of fuel spiked and we all sat on long lines at the gas stations and people began looking for ways to save energy. Those drafty old steam boilers that once burned coal and now burned oil had to go. Many people switched to natural gas in the early-70s, and boilers went in quickly because everyone was in a rush to save energy.

The contractors doing the fuel conversions usually sized the replacement gas-fired boiler based on the size of the nozzle in the old oil burner. Oil burner nozzles are rated in gallons per hour, and as I mentioned before, one gallon of fuel oil will produce 140,000 British thermal units of heat.

But keep in mind how that oil burner got sized as you mull over all of this. And consider, too, that the gasman is also going to add a generous safety factor when he does his sizing. He's concerned because the modern steam boilers were much smaller than the beasts they were replacing. That's part of what made them more efficient. They didn't contain as much metal or as much water. The gasman worried that he might not have enough water in the boiler to keep the boiler running while waiting for the condensate to return from the system. So he usually bumped up the boiler by at least one size.

Okay, let's go over this one more time. We begin with a boiler that can heat the house with the windows open and, if necessary, the roof removed. We nearly double that size because of the husband. And then the oil guy enters the picture and tosses in his safety factors. And then comes the gasman who does the same.

Which brings us to the 21st Century. It's probably time to replace that boiler again. Maybe that's why you're reading this book.

So, how do you feel about having a knucklehead size your new boiler based on what's in your basement right now?

I'd feel the same way.

Pros measure radiators, and look closely at the piping, before giving a price. That's the *only* correct way to size a replacement steam boiler. Don't buy more boiler than you need.

How knowledgeable pros size replacement steam boilers

It's not all that complicated. The boiler's ability to produce steam has to match the piping and radiation's ability to condense steam. That means someone with the right books has to go measure the radiation and the piping. (A really good book for this is *E.D.R. - Ratings for Every Darn Radiator (and convector) you'll probably ever see* compiled by yours truly. (You'll find it in the Books & More section of the HeatingHelp.com website.)

We rate radiators with a term called Square Feet of **E.D.R.** That acronym is short for **E**quivalent **D**irect **R**adiation. One Square Foot E.D.R. will emit 240 British Thermal Units per hour when the air in the room is 70° Fahrenheit and there's about one psi pressure inside the radiator. Any radiator's E.D.R. rating will depend on its size and shape and you need a good rating book to know what you're looking at. A knowledgeable steam heating professional will have one or more of these books. He will spend time going from room to room in your home with a ruler and a writing tablet. Each radiator will reveal its own rating and the pro will add the numbers together to get the final tally. If your contractor doesn't do this, then you're going to pay the price.

Once he has the total E.D.R from the radiators, the pro will add something to the total load for the piping. Your boiler has to bring all that pipe up to temperature on every heating cycle. The pro will look over the whole job and then base his piping factor on the size of the pipe, whether or not it's insulated, and on any radiators that are missing. Looking for missing radiators is important because the original job had pipe sized sufficiently large to deliver steam to all those radiators. If a homeowner has had radiators removed, either for decorating purposes or simply because the house was too hot, the piping will now be larger than it has to be for the current attached radiation load. The knowledgeable pro will compensate for this by slightly increasing the factor he uses for the piping load. For example, the standard piping pick-up factor for steam heating is 1.33. If the pro notices missing radiation, he might use a 1.5 pickup factor for the piping. That can make a big difference when it comes to your new boiler's performance.

If a boiler is oversized, it will short-cycle and waste fuel. Short-cycling also shortens the life of the controls. If the boiler is undersized, it will eventually heat your home, but it will cost you more money in fuel to operate than it would cost to operate a properly sized boiler. To understand why this is so we have to back up for a moment. Remember how we talked about how steam is a gas that desperately wants to turn back into a liquid? And to do so, it will give up its latent heat energy to *anything* that's colder than it is. And once the latent heat is gone, you no longer have steam. And that's when your house stops heating.

This is why I'm telling you that to be properly sized, a steam boiler's ability to produce steam *must* match the piping and radiation's ability to condense steam. That's why a pro will spend so much time measuring your radiators when it's time to replace that old boiler. He needs to get an accurate EDR rating for the whole place.

Okay, earlier I explained how boiler manufacturers rate their boilers. First, they show the Input rating. That's the amount of heat that's coming off the fire and entering the belly of the boiler. Right after Input, you'll see the Gross rating. Sometimes the manufacturer will substitute the term, D.O.E. Heating Capacity for Gross, but these two terms mean the same thing. The Gross rating is the amount of heat that's left over after the boiler loses some heat up the chimney and through its jacket. The difference between Input and Gross indicates the boiler's combustion efficiency.

The next rating on the chart is Net. That's the actual radiation load, usually expressed in Square Feet EDR. If you take the boiler's Net rating and multiply it by a factor of 1.33, you'll get the Gross rating. In other words, the manufacturer is assuming that the amount of steam the boiler will need to heat the pipes that lead to your radiators will be about one-third of the total radiation load. When a pro sizes a replacement steam boiler, he figures both the piping and radiation loads because both piping and radiation have the ability to condense steam.

Think of it this way. Suppose you were to remove all the radiation and piping from a building and weigh it on an industrial scale. Let's say you come up with a total of 5,000 pounds of iron and steel. In sizing that replacement steam boiler, a pro's job is to bring that 5,000 pounds of metal from room temperature to steam temperature, which, in this case, is 215° F.

Can you see this in your mind's eye? You're going to need a certain amount of heat to get the job done. How much depends on the Specific Heat of the metal, but let's not complicate this. Right now we're talking about an undersized boiler and why it cost more money to operate than a properly sized boiler, one that's matched to the connected load. Just focus on that 1.33 pick-up factor for a moment. This factor represents the difference between the boiler's Gross rating and its Net rating. That's the amount of steam the piping will condense when you first start the boiler. But once the piping gets up to temperature, the pick-up factor drops out of the equation. To understand why this is true, it helps to imagine the pick-up factor as the starting winding in an electric motor. That winding is there to give the motor the torque it needs to get its rotor spinning. But once it's up and running, the motor's centrifugal switch drops the starting winding out of the play. If you put an electric meter on a motor and then start the motor you'll see a surge in power at the instant it starts, a quick blip that disappears as the motor comes up to speed. That's the starting winding doing its job. A similar thing happens in a steam system when the pick-up factor is at work. It heats the pipes on startup, but once it completes that job it's no longer needed. The pick-up factor is a steam system's "starting torque."

And this is how an undersized steam boiler can eventually heat a house. The burner starts and runs for a *long* time. During that time, your family members who are in the rooms furthest from your boiler will notice that their radiators aren't getting hot as quickly as the radiators in the rooms that are closer to the boiler. This is because the system has the ability to condense more steam (on start-up) than the boiler can produce. And the problem is especially noticeable during the spring and the fall because this is when the weather is relatively mild. The system starts and stops more often. The pipes don't stay as hot as they do during the dead of winter.

You may decide to move your thermostat to the coldest room (or someone might suggest this). Moving the thermostat will make the burner run even longer. This, of course, wastes fuel because someone in your house is bound to open the windows in the overheated rooms. And as the burner runs longer, the steam eventually manages to heat all the piping. That takes the pick-up factor out of the equation. Suddenly, the boiler is large enough to heat all the radiators!

The end result is that an undersized steam boiler will probably heat your home, but at a greater cost because the burner will run longer. A properly sized boiler and burner saves fuel, and a properly sized burner is one that matches the firing rate to the connected piping-and-radiation load. If you're living with an undersized steam boiler you may not even realize that it's costing you more than it should because you have nothing with which to compare.

Another good reason to find yourself a knowledgeable pro.

And please keep this in mind when you meet the low-bidder. He may be the guy with the undersized boiler.

"Hey, it's steam! Wadda ya expect?"

The piping around your boiler

With rare exception, new steam heating systems haven't gone into homes since the 1940s. Hot water heating and warm air heating nudged steam aside when we started building tract housing in the suburbs. Steam became something to be maintained, and in many cases, loved.

During the 1960s, most of the Dead Men who had grown up designing and installing steam systems had retired. These people were like old professors but they had very few students because of the Vietnam War. The young people were either off to war, or off to college and the old professors retired and died. We lost much of the anecdotal and empirical knowledge of steam heating that the old might have passed on by word of mouth to the young, but it wasn't gone for good; it was still in the old books.

As we lived through the oil embargoes of the 1970s and rising fuel prices of the 1980s, many heating professionals went back to school to learn about steam heating and we're in much better shape today than we were just a decade ago. A lot of young professionals have fallen in love with the history and the mystery of the steam, and that's good news for you. We need people who can figure out what needs to be done, and to do it without ripping your house apart, or telling you that it's impossible. Always remember that it's mechanical, and that it can be learned, and there are many who *have* learned.

Point being, you don't have to be ancient to be good at solving steam heating problems. You just have to study and follow directions, and this is especially true when it comes to piping that new steam boiler. With the new boilers came new ways to pipe them, and that piping becomes critical to the replacement boiler's operation. The piping around the boiler (what we call **near-boiler piping**) has literally become a part of the boiler. Unless it's done properly, the new boiler won't perform well. The steam will condense before it has a chance to reach your radiators, and you'll be miserable.

So here are the key things to watch out for.

See those risers that leave the boiler and connect to the horizontal header above the boiler? That header must be at least 24 inches above the centerline of the boiler's gauge glass. This helps to dry the steam and every boiler manufacturer insists on this.

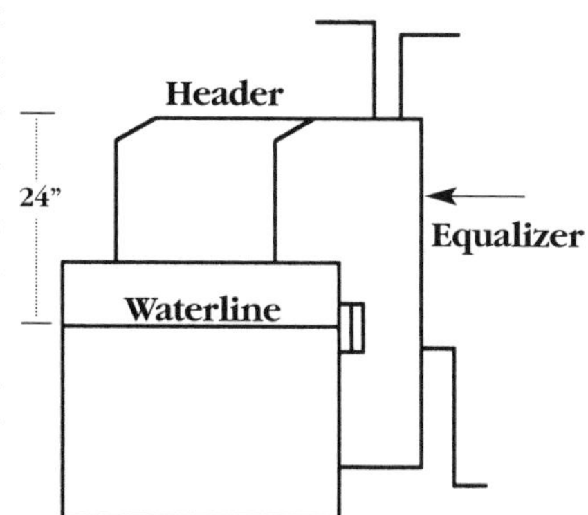

In addition, some boilers require more than one riser to the header. The more risers there are, the slower the steam will leave the boiler. There's a critical velocity (based on pipe size) that the steam can't exceed. If it does, the steam will drag boiler water with it as it leaves. The manufacturer will determine this velocity and note in their installation manual how many risers are required to keep the steam's speed within the limit. Ask your contractor to show you this manual.

The horizontal header also has to be the right size. That size will vary based on the size of the boiler (the larger the boiler, the wider the header), but the header should *never* be smaller than two-inches in diameter. The length of the header is not critical; it's the diameter that matters. Again, ask your contractor to show you the boiler manufacturer's installation instructions and talk to him about the proper header size. If your contractor wants to go larger on this size, that's fine. In fact, it's better. The wider the header is, the drier the steam leaving your boiler will be, and that's good because dry steam leads to lower fuel bills and fewer system problems.

Near the end of the header is the equalizer. This pipe does two things: It allows any water that makes it into the header to drain back into the boiler, and it applies pressure to the return side of the boiler, which helps to keep water from backing out of the boiler when the boiler is making steam. I'll tell you more about this in a minute when we talk about the Hartford Loop. What I want to mention right now, though (and this is real important), is that if the boiler has two risers to the header, the takeoffs that leave the header and connect to the steam mains, *must* come from a point that's between the last riser to the header and the equalizer. Like this.

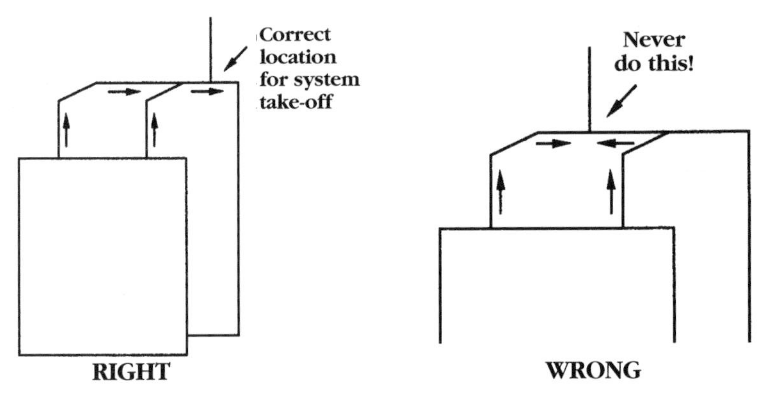

The steam will be leaving your boiler at very high speed and if it should pull some water from the boiler you want to make sure that that water heads for the equalizer and not for the radiators. Once it's in the equalizer it can get back into the boiler. If it gets into the radiators you'll have problems (spitting vents and possible water hammer). By positioning the takeoffs to the steam mains as I've shown them in the RIGHT drawing, everything will be heading in the same direction. The takeoffs can snatch the steam out of the flow while the heavier water sails by and drains into the equalizer. The WRONG drawing has the steam and the carried-over water meeting in a head-on collision. There's no way that the water can work its way backwards against that onrush of steam. The water winds up in the system and you're left with the headaches.

The low-bidder may have the WRONG piping diagram in mind when he gives you that attractive price. It's easier to pipe a replacement boiler the wrong way than it is to pipe it correctly, and that's why his price looks so tasty. Ask to see the manufacturer's installation drawings *before* you sign the contract, and have the contractor write into the contract that he will install your new boiler in accordance with the manufacturer's instructions. This is in your self-interest because if there's a problem afterwards, the boiler manufacturer may not honor their warranty if the boiler is installed incorrectly. You'll then find yourself between the contractor (who tells you that the manufacturer doesn't know what he's talking about) and the manufacturer (who will say the same about the contractor). If the contractor refuses to write into his contract that he will install the boiler in accordance with the manufacturer's installation instructions, find yourself a new contractor. And if he tells you that he's been installing steam boilers all his life and that these manufacturers nowadays don't know beans about installation, and that his way is fine and that he's never had a problem, not once, know that this guy is a knucklehead.

Know, too, that the near-boiler piping that's serving the old boiler probably won't work with the replacement boiler. New boilers are smaller and have special needs when it comes to this piping. If the contractor tells you that the old piping worked fine for all these years, and that it will work just fine with the new boiler, and that the old piping is like an old friend, and that he can do the job cheaper if you let him use it, know you are listening to someone who has not done his homework.

And if he tells you that copper pipe is better to use on the near-boiler piping, and that copper will save you money, tell him goodbye. Copper piping often comes apart after a few years of trying to keep up with the twisting that goes on within the near-boiler piping. By then, he's long gone and you're left with the problem. You'll have to find someone to fix it, which means you're going to pay twice for one installation. That guy's not really the low-bidder. The other part of his price (usually the most expensive part) is waiting for you down the road.

Copper also leaches out of the pipe and enters the cast

iron boiler, where it can cause dielectric corrosion and shorten the life of your new boiler. Is copper cheaper to install? You bet it is. And that's why your knucklehead is suggesting it. Insist on steel pipe.

And all that near-boiler piping must be insulated. There's no way around this. A professional will do this automatically, but have him write it into the contract anyway. He'll respect you for that.

When it comes to replacing a steam boiler, you get what you pay for. The low-bidder is usually the guy who will pipe the boiler wrong every time and then tell you that the problems you're experiencing afterwards have nothing to do with him. He'll swear that these boiler manufacturers don't know what they're doing, and that he has more experience than they do and that the heat in your home is uneven and the pipes are noisy because that's just the way steam heating is. And then he'll blow you off.

Ask to see those installation instructions before you sign the contract. Ask the contractor to write in the contract that he'll follow the manufacturer's installation instructions to the letter. Most low-bidders will run away at that point.

Let them.

The Hartford Loop

The Hartford Loop is a special piping arrangement that brings your boiler's equalizer and its return piping together. Follow along on this drawing as I explain it to you.

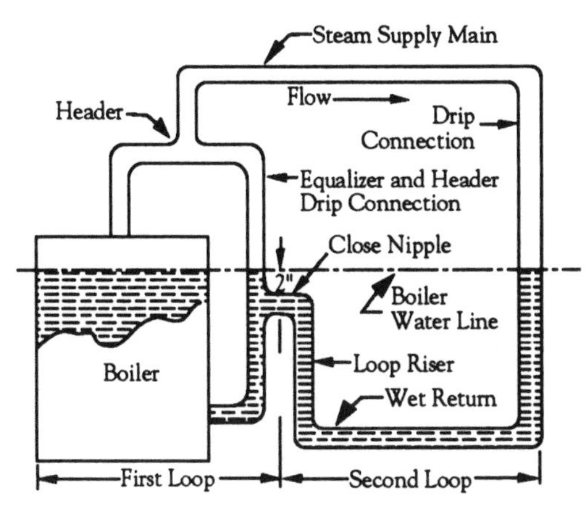

Okay, notice the way the end of the boiler's header pipe drops vertically below the boiler's waterline and connects into the bottom of the boiler. As I mentioned, we call that vertical pipe the **equalizer** because it balances the pressures between the boiler's steam outlet and the condensate-return line near the bottom of the boiler. The "wet" gravity return line, which returns the condensate from the system, rises up from that low point near your basement floor to join with the equalizer at a point that's about two inches below the boiler's normal waterline.

The Dead Men didn't always use this piping arrangement. There was a time when they would bring the return directly back into the bottom of the boiler without benefit of either a Hartford Loop or an equalizer. Like this.

When they piped a boiler this way, the slightest steam pressure inside the boiler would push water out of the boiler and into the condensate-return line. That presented a problem because if enough water left the boiler, the boiler would overheat and that's not healthy for the folks in the building. So to keep the water from backing out, the Dead Men would use a check valve in the wet return (the pipe below the boiler waterline). You'll recall that a check valve is like a turnstile.

It lets water go one way but not the other. Here's where they put it.

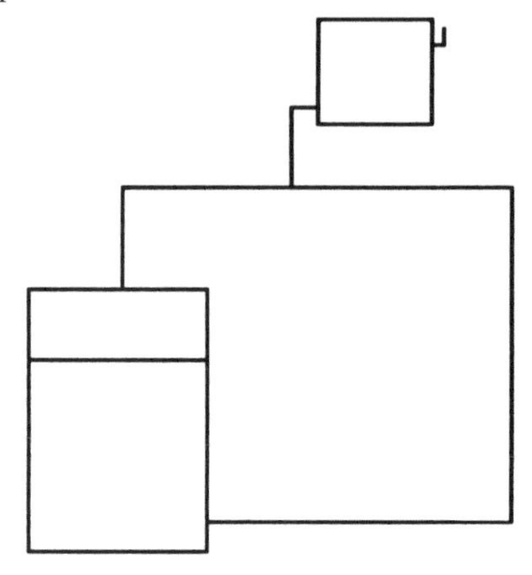

Before long, though, the Dead Men learned that the check valve was susceptible to the slightest amount of crud in the system. The flapper would get stuck open and they'd be right back where they started, with water backing out of the boiler and danger lurking.

So they put their heads together and came up with the idea of an equalizer pipe to replace the check valve. It works like a charm. Whatever pressure is inside the boiler will also be inside the equalizer pipe. The two forces balance each other, and the water stays in the boiler. Here's how that looked.

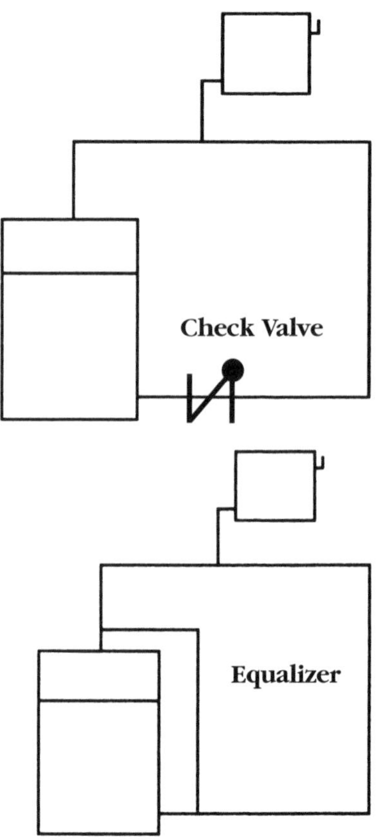

Okay, here comes the scary part (and the Dead Men lived with this every day). If a return line broke, water would flow from the boiler, and the boiler would either crack or explode. Some choice, eh? There was nothing to prevent this from happening because there were no low-water cutoffs in the days of coal-fired boilers. Low-water cutoffs didn't show up until 1923, and they needed either a gas burner or an oil burner to work. I'll tell you more about them in just a little while.

And that brings us to this piping arrangement that came to be known as the Hartford Loop. Around 1919, the Hartford Steam Boiler Insurance and Inspection Company got tired of paying the claims on all those cracked (and exploding) boilers, so they came up with the idea of this special piping configuration that would be pretty foolproof. They wouldn't insure a boiler unless it had one. Before long, everyone in the trade was calling it the Hartford Loop, or the Underwriters Loop, both after the Hartford Insurance Company. It worked beautifully and if you were to look at the records of boiler failures before and after 1919, you would see that the Loop really made the world a safer place.

And its operation is so simple. If a return line breaks, water can only flow from the boiler to the point where the wet return line connects with the equalizer. The Loop works like a siphon that runs out of water (it's like the line you'd use to drain a swimming pool). The point where the Loop connects to the equalizer is slightly higher than the boiler's crown sheet (that's the internal metal "roof" of the boiler's combustion chamber, the part of the boiler that contains the fire). If a return breaks, water can only drain from the boiler to that level, and that level is still safe. The Loop gave the Dead Men a chance to notice the problem and to act. It wasn't infallible, of course, but it was a vast improvement over what was there before.

Nowadays, boilers have low-water cutoffs to watch the waterline and shut off the burner should the water level drop too low. So do you still need a Hartford Loop? You bet! A Hartford Loop is the cheapest insurance you can have on hand to back up that low-water cutoff should a return rupture and water suddenly leave the boiler. Low-water cutoffs sometimes fail, and having a Hartford Loop in place is like wearing a belt and suspenders.

Now, notice how the Loop connects into the equalizer with a very short piece of pipe.

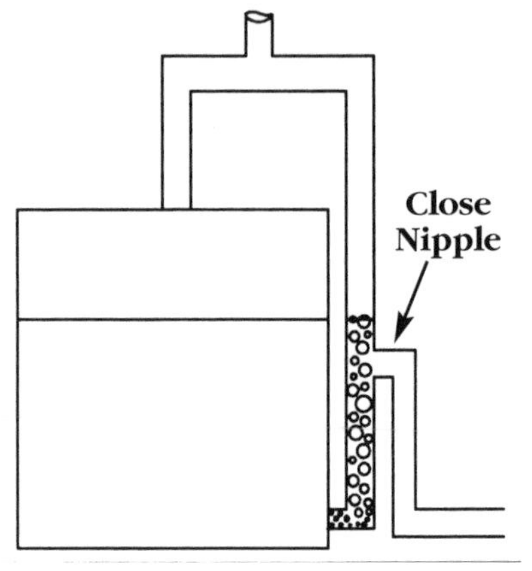

That pipe is appropriately called a **close nipple** and it serves an important purpose. Steam is rising through the water that's inside the equalizer, just as it rises through the water inside the boiler. Keep in mind that steam takes up nearly 1,700 times more space than water so when those big steam bubbles meet the relatively cool condensate that's returning from your radiators (through the Hartford Loop), the steam bubbles are going to lose. They'll instantly collapse and the returning water will rush in to fill the void. That rushing water can create water hammer when it makes contact with the back wall of the tee fitting that connects to your equalizer. The longer the horizontal connection between Loop and equalizer, the greater the potential for water hammer, and that's why the Dead Men (and savvy installers nowadays) will use either a close nipple for this important connection or a Y-fitting (which looks like that letter). A long nipple will act like a gun barrel for the returning condensate. Not good.

If you're getting a new boiler, the installer should check the manufacturer's specs on the height of the connection to the Hartford Loop. If the connection is too high you can get water hammer. If it's too low your boiler won't be protected.

If you have one-pipe steam, a good steam-heating installer will know that the Hartford Loop has to connect to the equalizer like this.

In a one-pipe system, the end of the steam main is also the end of the condensate return line. In this system, the main goes all the way around the system and stays above the boiler's waterline every step of the way, which might make you wonder about the purpose of the Hartford Loop in this case. There are no return pipes below the boiler's waterline, so there's really no danger of losing the boiler water should an above-the-waterline return pipe break.

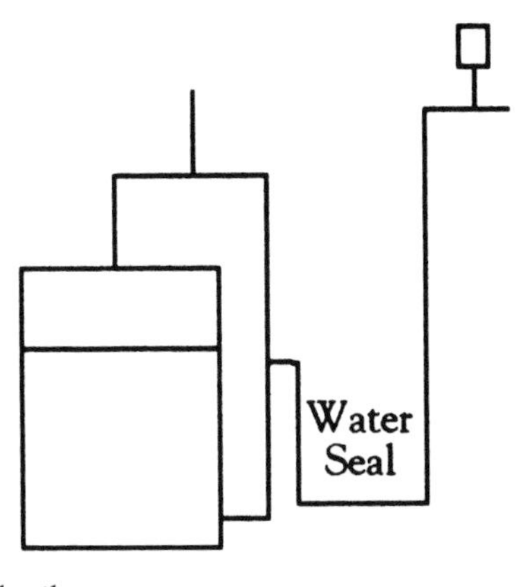

But keep in mind that the boiler still needs that equalizer to keep the water from backing out of the boiler when steam pressure builds. Suppose the end of the main were to drop directly into the middle of your equalizer like this.

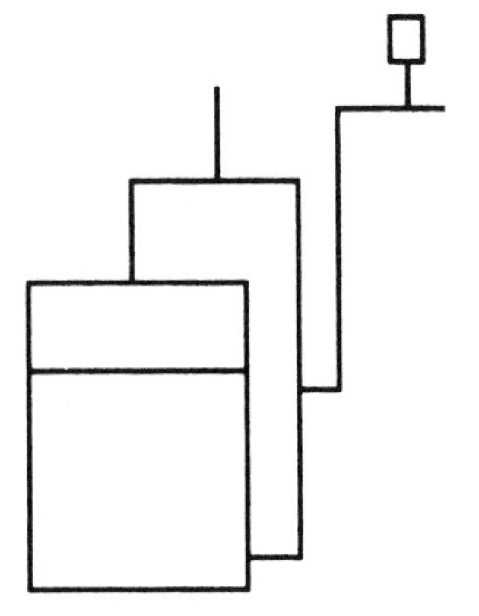

Can you see how you'd be setting up a condition where steam might have access to the return line through the equalizer as the waterline in the boiler steams down. By bringing your return line down to the floor, and then rising up into your equalizer (in other words, by building a Hartford Loop), your installer will be making sure that steam in the equalizer can never enter the return line and cause water hammer.

What size pipe should the installer use for the equalizer? Well, that depends on the size of your boiler. Boilers that have a D.O.E. Heating Capacity (or Gross rating) up to 216,000 Btuh (that's a big house!) should have an equalizer that's at least 1-1/2" in diameter. If your boiler is rated up to 1,500,000 Btuh, you're living in a mansion and you deserve a 2-1/2" equalizer. Anything larger than that means that you're rich enough to have someone else read this book for you, so don't worry about it.

If the equalizer is too small it won't balance the pressure on the return, and your boiler's waterline will be very unsteady. Never accept an equalizer that's smaller than 1-1/2". And I don't care what the guy tells you. Don't go for it.

If your heating system has a condensate return pump (to return the condensate to the boiler), then you don't need a Hartford Loop.

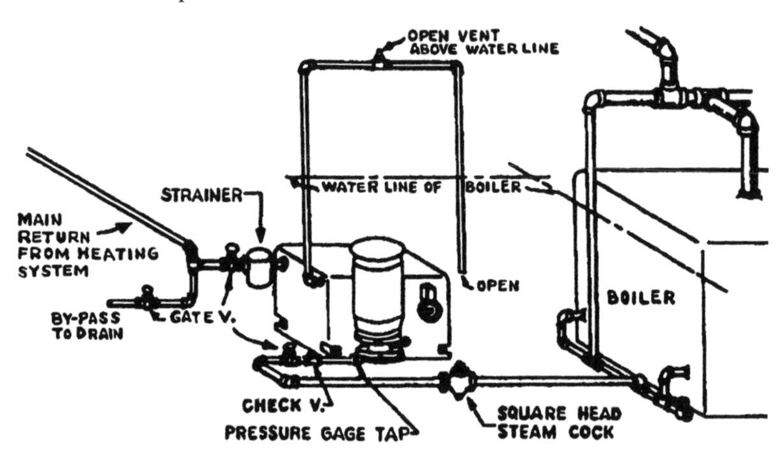

The condensate pump is vented and it opens the return side of your system to the atmosphere. At that point, the equalizer stops being an equalizer, and becomes just a drip line for the boiler's steam header. All the returning water must flow into the pump's receiver, and from there, into the boiler. The pump has a check valve at its discharge to keep the boiler water from flowing out of the boiler. Should the check valve fail, the water from the boiler will simply flow backwards into the pump's receiver, and start the pump. The pump will move the water back in the boiler and then shut off. Then it will do it again – over and over.

Should the receiver spring a leak *and* the check valve fail at the same time, it's possible for the boiler to lose its water, so you can insist on a Hartford Loop, but only if you like to wear two belts and five pair of suspenders. There is a drawback to pumping into a Hartford Loop, however. When the pump starts, it will shoot water under pressure through that close nipple and into the back of that tee that connects the Loop to the header drip. Some of that water will probably fly up into the boiler header and leave you with the banging and clanging of water hammer.

You're better off if your pump discharges into the bottom of the header drip, well below the boiler's waterline.

Holey boilers!

George and Dolores live across the street from us. I was looking out the window one day in February and I noticed that there was white smoke coming from their chimney. A *lot* of it. I walked across the street and knocked on their door, not really wanting to be the bearer of bad news, but they needed to know. George came to the door and I started to point up at the chimney when he said, "Yeah, I know. They're changing it tomorrow."

The guys from the oil company showed up early the next morning and worked all day. Nosey neighbor that I am, I wandered over to look at the old boiler that sat in his driveway, waiting for its ride to the dump. It was in sad shape. It had a hole in it and water had been escaping into the combustion chamber. When their oil burner started, the flame turned the water to "white smoke" (actually wet steam) and that's what I had seen coming from their chimney. It looked very Currier and Ives from my side of the street, but it wasn't such a good day for George and Dolores, who had to replace their old boiler in a hurry.

Oxygen is the only thing that can rot a hole like that in a boiler. Oxygen enters the system by way of the water that we add to the boiler. Many steam boilers have automatic water feeders so no one notices that this is happening until it's too late. The oxygen comes out of solution as the water comes to a boil and it rots the metal right at the boiler's waterline. Since the hole is right at, or slightly above the waterline, the boiler doesn't leak onto the floor. The water just boils off and heads for the chimney.

If this is happening to you, you'll probably notice it either from the Currier and Ives effect, or when your house doesn't heat as well as it used to. Everything was fine last year, but suddenly your fuel bills are higher than ever and half of your rooms are cold. That's because the steam that's supposed to go to your radiators is going up the chimney instead.

Most often, the reason the boiler is taking on so much fresh water is because there's a leak in a pipe. That pipe is usually the one that's buried under your basement floor. One of the ingredients in concrete is fly ash, which comes from coal-fired power plants. Of the 57 million metric tons of fly ash produced in the US each year, 17 million metric tons goes to the concrete industry. According to the National Ready Mixed Concrete Association, virtually every batch plant uses fly ash. It's cheap, plentiful and it works. And by the way, this is why they call *cinder* blocks by that name.

The cinders used in concrete are acidic and they will eventually damage pipes that are buried underground. The Dead Men who knew their stuff would put crushed limestone around the pipes before burying them in concrete. Limestone is alkaline so it balances out the acid in the cinders. But if no one added the limestone years ago, the pipe will slowly dissolve from the outside in.

But wait, there's more! (I love saying that.) When the water boils in your boiler it releases carbon dioxide. This comes from the breakdown of carbonates and bicarbonates in the feed water (a natural occurrence). The carbon dioxide gas mixes with the condensate on the return side of your system and forms a mild carbonic acid (similar to what's in Coke and Pepsi). That acid does a nice job of eating through the pipes from the inside out. It takes time, but it happens.

So you have those two chemical actions working on your pipes – carbonic acid and sulfuric acid, munching away from different sides and eager to meet up with each other. Murphy's Law states that it will usually be the hidden pipe that goes first, the pipe that's hardest to get to. And when it leaks, the water will flow downward into the ground, and the automatic feeder will go to work to replace the missing water so that the boiler can keep running. Which it does, until your boiler develops a hole through which you could toss a tomcat.

One bad main vent can also cause a large loss of water. This vent is usually the one that's out of sight in a crawl space. The crawl space fills with steam and no one notices that anything's wrong until the hole appears in the boiler.

Beyond the obvious damage to your level of comfort and your family's fuel budget, the scariest thing about having steam in your chimney is that your chimney contains products of combustion (a polite way of saying "soot"). When these get wet they form sulfuric acid that can eat the mortar that holds the chimney's bricks in place. As bricks loosen, the draft going up the chimney suffers and this can back deadly carbon monoxide into your home. (And just as an aside, if you don't own a good carbon monoxide detector, *please* get one today. It just might save your life. The one we use is from CO-Experts, www.coexperts.com. Good stuff.)

If you suspect that you may have a hole in your steam boiler, the simplest way to check is to open the feed valve and flood the boiler up into the header pipe. If there's a hole, water will pour from the boiler and onto your basement floor. At that point, it's time to replace the boiler.

And this doesn't just happen to old boilers. I've seen it happen to boilers that are not even two years old. It's the feed water that does it. Steam boilers need feed water, but too much of a good thing leads to trouble. System leaks lead to excessive feed water. Have the leaks fixed.

If you have an automatic water feeder, have a professional install a water meter on that line. They're not expensive and it's the best way I know to prevent holes in your boiler. Hang a small notebook on the water meter and keep track of its reading. Do this once a week when you blow down your low-water cutoff (more on low-water cutoffs coming up). If you're thinking about adding an automatic water feeder to your system, check out the one offered by Hydrolevel Corp. (www.Hydrolevel.com). This unit has a built-in water meter that keeps track of how much water passes into your boiler. Check it once a week and you'll know immediately if your system has developed a leak. There's a picture of this unit in the section on electronic water feeders. Coming up!

If you find yourself with a holey boiler, please don't try to fix the hole by adding a stop-leak product, such as the ones used on automobile cooling systems. Those chemicals may find and stop a small leak, but they'll also find and seal your air vents and contaminate your boiler water to a point where it won't be able to produce quality steam. The water will be riding the steam up into your system, resulting in water hammer, uneven heat, high fuel bills and headaches for you. And that leak will redevelop sooner or later. It will probably wait until you're on vacation. Life's funny that way, isn't it?

Some of the Dead Men used oatmeal to repair small leaks. The idea was that the oatmeal would clump up at the hole. Other old timers used horse manure for the same purpose.

Would you want to rely on oatmeal and horse manure to fix a leaking boiler that's in the basement of your home?

I wouldn't either.

Which is why I think that if a boiler is leaking, it's time for that boiler to retire.

How "High" is High-Efficiency?

If you're replacing your old steam boiler you're probably looking for something that's more efficient than your old one. A fine idea! My guess is that your old boiler is sending about 40% of your fuel dollars up the chimney. When you're shopping around for a contractor you'll most likely ask them about boiler efficiency. Who doesn't want to save on fuel? But don't be surprised if the best combustion efficiency they tell you about is in the low-80% range. That's about as good as it gets with steam. Modern hot water boilers are more efficient than modern steam boilers because of the differences in the systems. A steam boiler has to bring the water temperature to 212°F. to make steam. A hot water boiler that's operating within a system that has, say, baseboard convectors can run cooler, typically at a high-limit of 180°F. A hot water system that incorporates radiant floors will be even more efficient; it might operate with water that hovers around 110°F. There are boilers used with radiant systems that run with a combustion efficiency that approaches 100%, and some of these boilers don't even need chimneys. The flue gases just condense to a liquid inside the boiler and flow through a small plastic pipe to a drain.

But with steam boilers, low-80% is what you're looking at. That's about as good as it's going to get.

And by the way, speaking of chimneys, this is something that your contractor definitely should look at (and if he's a pro he will). A modern boiler, one that operates at higher efficiency, can damage an old chimney because the flue gases that go up the chimney are at a relatively low temperature (compared to what left the old boiler) and close to their dew point. Dew point is the temperature at which a vapor begins to condense. That's what makes the boiler efficient; more of the heat gets transferred to the water so the flue gases that head for the chimney are cooler. If those flue gases condense in your old chimney they can form sulfuric acid, and that can weaken the mortar between the bricks and put you in danger of carbon monoxide poisoning should the bricks fall down and block the chimney.

Fortunately, unlike their hot water brethren, modern steam boilers are not superstars when it comes to high-efficiency ratings, so this is less likely to be a problem when you're having your old steam boiler replaced. However, if your contractor notices a weakness in your old chimney, and this is more likely if the chimney is outside your home (because it's colder there), or if the chimney has gone for years without a cap (which allows the rain to pour in), then he may recommend a stainless steel liner. This goes inside your old chimney, and although a stainless steel liner can be relatively expensive, it's there for your protection, so don't try to get by without it if a pro you trust insists on installing one with the new boiler. A real pro won't do the job at all if he can't do it right.

And let's face it, it's your life at stake. This is not the best place to be cutting corners.

And when it comes to efficiency, know that there's more involved than just the boiler. We're dealing with a whole system here, so think in terms of *system* efficiency and not just the boiler's combustion efficiency. You could have a brand-new boiler and I could show up and plug just one main air vent down at the end of the line. That one vent will slow the entire system, cost you money and make you uncomfortable. The same goes for a clogged wet return line, missing insulation on the pipes, poor boiler water quality, bad steam traps and many other things. Think *systems*, not just boilers when you're thinking efficiency.

A new, but oversized, boiler can waste fuel because it will short-cycle. A pro will size your boiler based on the connected piping-and-radiation load. There's no substitute for this. As I mentioned earlier, in the old days, they oversized steam boilers because of the need for open-window ventilation. Chances are, the boiler in your house is a lot larger than it needs to be. If you replace it with a new boiler of the same size without first having an audit of the connected piping and radiation done, you'll probably wind up with an oversized boiler. The efficiency of that new boiler will look great on paper, but it won't prove out as the years go by.

An undersized boiler will also waste money, as I mentioned

before. Efficient on paper, lousy in your basement. Have it sized right.

Also, a thermostat that's not set properly (and here I'm talking set internally) can also cause a boiler to short-cycle and waste fuel. Thermostats contain a device called an **anticipator**. Its job is to anticipate the setting that you've selected for your home and to shut off the burner before the room air temperature actually reaches that setting. By "anticipating" the coming warmth, the anticipator keeps the system from overshooting the mark and making your home too hot. It takes an electrical meter to properly adjust a thermostat's anticipator and the settings will vary from job to job.

Work only with a knowledgeable pro when it's time to replace that old boiler. That's the best way I know to deliver *system* efficiency.

Those Glorious Radiators!

They made old-fashioned steam radiators pretty much the same way they make boilers nowadays. They poured molten cast iron between two sand castings and they wound up with a radiator section that was hollow. They joined this to another, identical section, and then to another, and another, until they had a radiator of the right size for the room it would serve.

In the early days, they joined the radiator sections by connecting only the bottom portion of the sections. They used a fitting called a **nipple** to do this. The first nipples had left-hand and right-hand screw threads. These were appropriately named **screwed nipples**. The Dead Man would place the nipple between two sections of the radiator, and then he'd turn it with a special wrench. Since the threads were left-hand on one side and right-hand on the other, when he turned the nipple in one direction it drew the two sections together. If he turned the nipple in the opposite direction it would push the sections apart. As the years went by, though, these screw-type nipples rusted in place and there's really no way to get them out nowadays. If you have radiators with this type of nipple you won't be able to remove any of the sections, should you wish to make your radiator smaller, or should one of the sections spring a leak.

The other type of nipple (and this one came later) is a

push nipple. This one has no threads on it. It's wider in the center than it is on the edges.

The Dead Men would lubricate the push nipple, place it between two sections of the radiator and then bring the sections together by tightening tie rods that ran from one side of the radiator to the other. You *can* remove the sections of a radiator that has push nipples, but it requires pry bars, hammers, muscle, care, prayer and luck. And by the way, if you don't see the tie rod between the sections, you have screwed nipples on that radiator.

In the beginning of the 20th Century, the Dead Men began to use freestanding, cast-iron, hot water radiators on their steam jobs. The difference between a hot water radiator and a steam radiator is that the steam radiator is connected only across the lower part of the sections, and the sections (when viewed from the side) have **columns** that are fairly wide. Here's a picture of a steam radiator with three columns.

These terms can sometimes be confusing so let me explain it again. A **section** is an individual casting, a single slice in the "loaf of bread" that is the radiator. Push nipples connect the sections. A **column** is what you

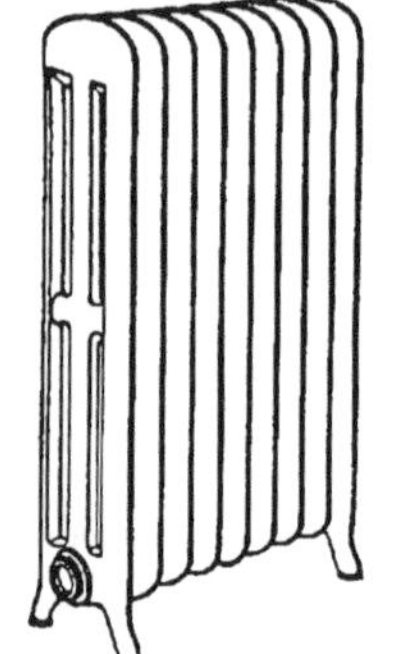

	Height (Inches)	Sq. Ft. Per Section
Three Column	18"	2.25
	22"	3
	26"	3.75
	32"	4.5
	38"	5
	45"	6

see when you look at a single section from its side. Columns create air spaces in the sections so that air can reach in and pick up the heat off the iron.

When you want to know a radiator's heat output, you'll first look at the side and count the number of columns. Once you've determined that what you're looking at is, say, a three-column radiator, you'll then measure its height. Then you'll count the sections. Finally, using a chart such as the one above, you'll look up the radiators rating in Square Feet EDR.

Not too difficult, is it? The key lies in having the right reference books because there were hundreds of different types of radiators and the closer you can get to the actual load, the closer you can get to the proper boiler.

Now a hot water radiator is different. It will always have nipples across both the upper and lower parts of the sections. They make them that way so that water can circulate completely across the radiator. Here's a picture of a hot water radiator.

Viewed from the side, you'll notice that the divisions of each section are narrower than the columns on the steam radiators. We call these **tubes** instead of columns. This type of radiator was more modern looking. While today, we fuss over and try to restore and preserve Victorian radiators, most people in the Thirties and Forties thought these looked old-fashioned. That's what led to the streamlined design of the tube radiator.

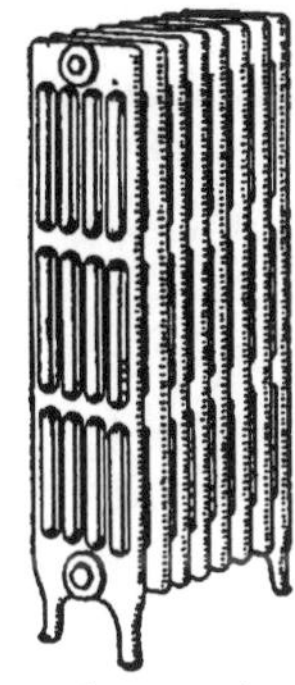

	Height (Inches)	Sq. Ft. Per Section
Five Tube	20"	2.66
	23"	3
	26"	3.5
	32"	4.33
	37"	5

A practical advantage came to light, though, when the Dead Men began to use these hot water radiators on steam. Those upper connections allowed them to supply the radiator from the top and to remove the condensate from the bottom. The steam worked its way across the top of the radiator, pushing the air downward and giving up its latent heat to the air in the room. At the same time, the condensate drained down the inside of the radiator, passing its sensible heat into the room as well. The result was a radiator that was more evenly hot and that resulted in a nice radiant glow. This type of radiator just felt better to be around, and that's one of the things that led to its popularity.

With the old-fashioned, column-type radiators, steam would enter through the bottom and rise to the top of the first section, displacing the air, which would then move over into the next section. The steam threaded its way up and down each section, as though it was lacing a sneaker. On milder days, this type of steam radiator probably won't get hot all the way across. The thermostat will shut off the burner before all the sections get hot. That's normal. Tube-type radiators often get hot all the way across because of the way the condensate drains along the inside surfaces.

And you can still buy brand-new, tube-type, cast-iron radiators. These companies make them.

- Burnham (www.burnham.com 717-397-4701)
- Governale (www.governaleindustries.com 718-272-2300).

And please don't worry about a fire hazard if paper comes in contact with your steam radiators. The surface temperature of those radiators gets to about 220°F tops. The flashpoint of paper is 451°F (remember the Ray Bradbury science fiction tale, *Fahrenheit 451?)*. So don't worry; your radiators might burn your skin if you touch them while they're hot, but they can't start a fire in your home, no matter how long they're on.

Which reminds me of this guy who called to ask a question about a house he and his wife were thinking of buying. "It's an old beauty," he said, "but it has steam heat."

"So, what's wrong with that?" I asked.

"Well, we have a small child."

"That's nice," I said.

"I mean, we're concerned that our child might touch a radiator and get burned. Did you have steam heat when you were growing up?"

"Sure did!" I said.

"And did you ever get burned?"

"Yep," I said.

"How often?" he asked.

"Just once," I said.

So don't worry.

How we size radiators

James Watt, famous for the invention of the steam engine, also invented the first steam radiator. The British had been using steam to heat their greenhouses for years, but they didn't use radiators (they couldn't because Mr. Watt hadn't invented them yet). They just had the steam spew from the end of a pipe. It would both heat and humidify the air in their greenhouses, and most everyone was pleased with that.

Everyone except Mr. Watt, that is. He figured he could warm his office with steam, but he didn't want his papers to get mushy so he ran the steam pipe to a metal box that sat on the floor of his office. When he wanted heat, he'd open a little valve on the box to let out the air. This became the world's first steam radiator, and from this came the term **Square Foot of Radiation**, which was literally one square foot on a metal box filled with steam.

As time went by, radiators got fancier. Check out this one, for instance.

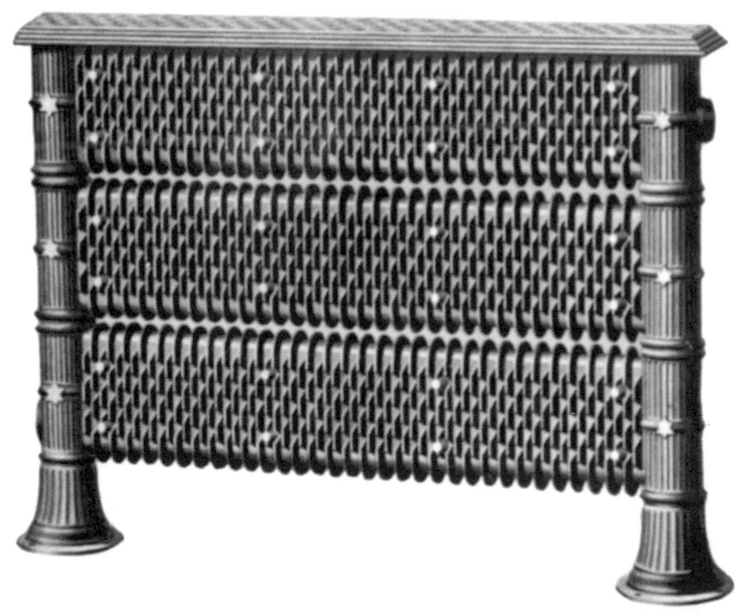

Isn't that gorgeous? When radiators such as this one

started showing up the Dead Men faced a real challenge. How do you measure the square footage on the surface of something that has so many nooks and crannies? They thought about it for a while and came up with an ingenious solution. They got themselves a large vat of paint, which they weighed on a scale. Then they sealed the radiator's inlet and outlet openings and lowered it into the vat of paint until it was completely submerged. When they raised the radiator out of the paint they let the excess drip off and then they weighed the vat of paint again. Whatever was missing from the vat was now clinging to the outside surface of the radiator. Pretty clever, eh? Then they took that much paint (the missing amount) and used it to paint the floor. However many square feet of floor they covered with the paint became the square-foot rating of the radiator.

Now, seriously, would you have thought of that?

They started to call this **Equivalent Direct Radiation** or **EDR** for short. Then they assigned it a real value. For steam, one square foot of EDR will give off 240 Btus per hour when there is 70°F. air on the outside of the radiator and 215°F. steam on the inside of the radiator. That would be steam at about one pound per square inch pressure.

Hot water radiators also have EDR ratings, but the temperature of the water in a hot water system is usually much lower than 215 degrees, so the value of EDR for these systems varies. For instance, a typical hot water heating system will contain water at an average temperature of 170°F., so the value of EDR will be only 150 Btus per square foot. That's something to consider if you're thinking about converting your steam system to a hot water system. The radiators might not be large enough to get the job done with hot water. More on this later.

When you hire a heating contractor to replace your steam boiler, he will go from room to room, measuring all your radiators. He'll then use a reference book to find the total square foot EDR for your system. If he doesn't do this, he is (you guessed it!) a knucklehead.

Pitch

One-pipe steam systems have just one pipe at each radiator and the steam and the condensate have to pass each other whenever the heat is on. For the condensate to drain from the radiator, the radiator has to pitch backward toward the supply valve. Common sense, right? Water flows downhill. If the radiator isn't pitched then the water will just build up inside the radiator and probably make a racket when the steam arrives, or it just might decide to squirt out of the air vent and make a mess on your walls and curtains.

The Dead Man who installed your radiators would have made sure each pitched properly. He would have used a shim under each radiator leg opposite the supply valve. But as time went by, those shims might have worked their way loose and your radiators may be level now, or even pitched in the wrong direction. A radiator is heavy and it's always expanding and contracting. Those iron legs opposite the supply valve can trench themselves down into the floor over time, especially if the floor is made of wood.

So it pays to check the level of your radiators. Do this during the winter when the system is running. The pipes will be hot and fully expanded. That makes a difference because, depending on its height, an expanded pipe can lift a radiator from the supply valve's side. Use a six-inch level and start from the side with the supply valve. Check the pitch every six inches. This is important because a large radiator can sag in the middle as it gets old, causing condensate to collect at the low point.

It doesn't take much pitch to cause water to flow from the radiator. Let your common sense be your guide here. Check the level and ask yourself, if I were water, would I be able to flow. Sounds silly, I know, but it's really no more complicated than that. If you need to pitch the radiator (and if you're feeling strong), use a lever and a fulcrum to *gently* (and I mean that) raise the free end of the radiator and then place something under the two feet on the side that you're lifting. A small block of wood works well, as do plastic checkers. In fact, checkers are perfect for this job because you can stack them and they're usually just the right diameter for the radiator's cast iron feet. And just imagine; you get your choice of red or black! Feel free to make a fashion statement.

And take care not to pitch your radiators too much because you're liable to snap a pipe. There's probably also a lot of collected rust and other goop at the bottom of your radiators. If you pitch them too much that goop might slosh toward the supply valve and form an internal weir behind which condensate will collect. That usually results in knocking and spitting from the radiator vent.

And in case you're wondering, two-pipe radiators can be level. Here, the outlet is at the bottom of the radiator and the supply valve is almost always at the top. The condensate builds up in the bottom of the radiator and the steam pushes it through the radiator's outlet, so pitch isn't critical here.

But back to one-pipe.

Where does the radiator air vent belong?

On a freestanding, one-pipe steam radiator, you'll have an angle-type air vent. These are the kind that have the screw connection on the side. Like these.

The correct place for the air vent is on the side of the radiator that's opposite the supply valve, and about halfway down that last radiator section. This is important because, as you know, steam is lighter than air. When it enters your radiator it's heading for the top. If the air vent is at the top of that last radiator section, rather than halfway down that section, the steam will shut the vent before most of the air has had a chance to get out of the radiator. You'll wind up with a room that doesn't heat well. So the vent belongs where you see it in this drawing.

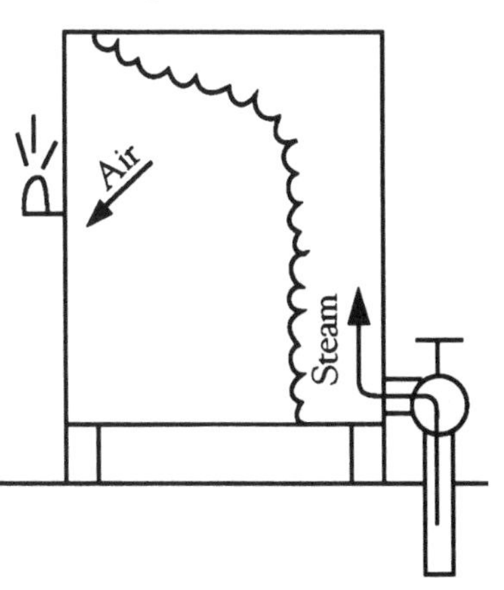

If it's not in the proper place on your radiators, look for a raised, round surface on the radiator in the spot where the vent should be (as in the picture). That raised surface is called a **boss**. A pro can drill and tap that boss for a radiator vent. This takes special tools and is not something you should try to do on your own. On the other hand, the boss may already be drilled and tapped and there might be a pipe plug in it. If that's the case, remove the plug and put the vent there. Use the plug that you took out and screw it into the hole where the vent used to be.

If you have steam convectors such as the one I'm showing you here the vent will be a bit different.

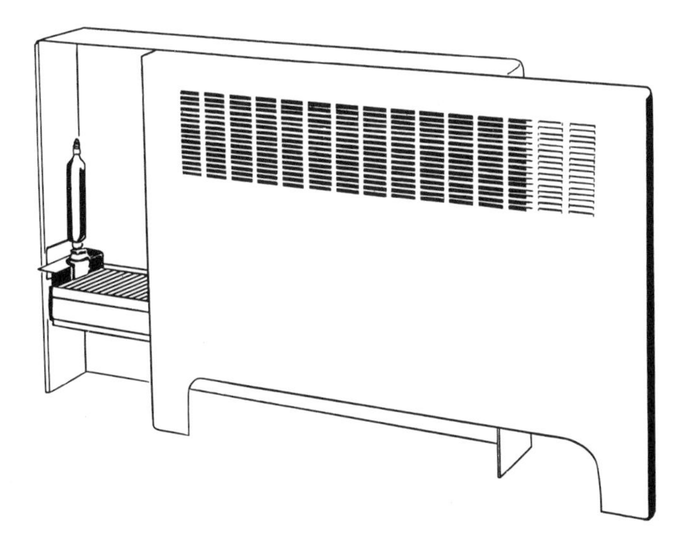

If you remove the front metal cover on the convector you'll see a supply valve and a horizontal element, which might be made of cast iron, steel, copper, or a mixture of these metals. If it's a one-pipe radiator, the air vent belongs on the side opposite the supply valve and it's going to have a screw connection coming out of its bottom rather than its side. Like one of these.

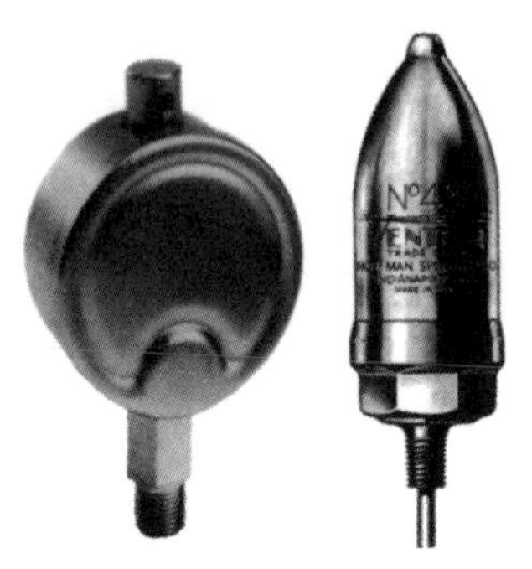

Air vents have no business being on two-pipe steam radiators. These radiators have steam traps and an air vent can hide the symptoms of a failed steam trap, causing you big problems down the road as water hammer tears your system apart. Never let anyone add air vents to your two-pipe radiators.

Other types of radiators

The classic steam radiator is the cast iron type that sits heavily on the floor. Typically, that radiator will be beneath a window because that's usually where the most heat is lost. Ideally, the radiator should be 2-1/2 inches away from the wall and its top shouldn't extend above the windowsill. A floor-mounted radiator can be anywhere in the room, however, and still get the job done. The Dead Men sometimes mounted the radiators in places other than under the windows for piping convenience, or the homeowner may have requested this.

Sometimes, they even hung radiators on the wall, a few feet up from the floor. Or they'd screw them into ceiling. In both cases, this was because the radiator was on the same floor as the boiler. Raising the radiator puts it above the boiler's waterline. That ensures that there's steam and not water inside the radiator.

Convectors

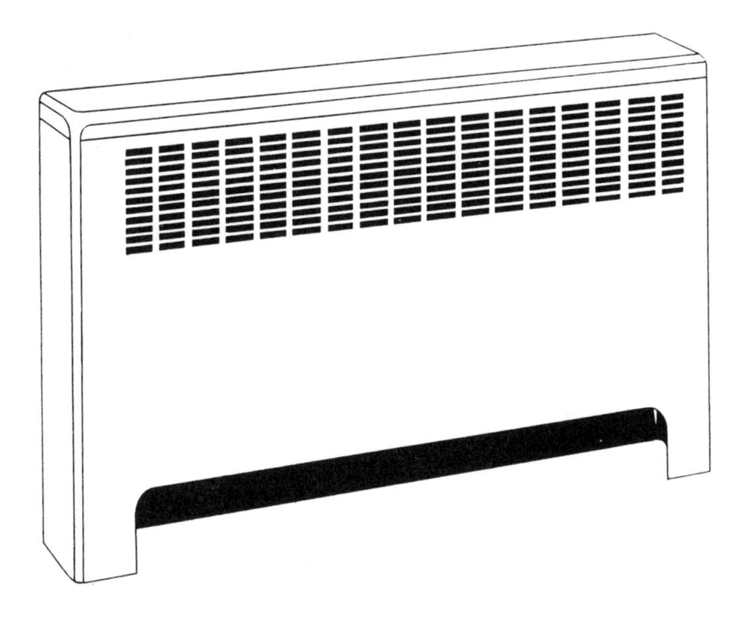

Convectors heat air by convection (hence the name) and the cabinet design is so important. The coolest air in the room is also the heaviest air, so it's going to stay near the floor. Inside the convector we have a heating element that's made of steel, copper, cast iron or a combination of these metals. When steam enters the heating element it gives up its heat to the surrounding air, causing that air to rise within the convector cabinet, and leave through the holes near the top of the cabinet. As the heated air rises, it creates a chimney effect within the cabinet, which draws in more cool air from the floor. As this continues, an invisible Ferris wheel of rising and falling air forms inside your room. You may notice that the walls directly above your convectors sometimes get streaked with dirt. And that's exactly what that is. It's dirt. Even though you keep a clean house, tiny particles of dirt will ride the invisible Ferris wheel of convective air and get stuck to your walls. This is especially true if someone in your house smokes, or likes to burn candles or incense.

If you find that a room with a steam convector is too hot for your tastes, a quick fix is to cover a portion (or all) of the convector's heating element with aluminum foil. The foil will slow the air that's trying to rise from the heated element and this is often all it takes to calm down that overactive convector. Another option is to use a thermostatic radiator valve. This either replaces the supply valve on a two-pipe steam system, or fits between the air vent and the convector element on a one-pipe steam system. The thermostatic radiator valve is self-contained and non-electric. It senses the air temperature in the room and either stops the steam from entering the convector element (on two-pipe systems), or keeps the air from leaving the air vent (on one-pipe). It's best to hire a pro to do the two-pipe work. If you have one-pipe steam, and you're feeling handy, you might try to do this project yourself. Make sure you get the type of thermostatic radiator valve that has a remote sensor, though. You don't want it to sense the air temperature within the convector cabinet. I'll tell you more about thermostatic radiator valves in a little while.

Recessed cast-iron convector

Many homes have these recessed units. They're on both one- and two-pipe systems and the nice thing about them is that they sit back and out of the way of the people. They deliver wonderful radiant heat because they look out at the room from their little caves with a full face of cast iron. And they also heat beautifully by convection. It's like having the best of both worlds – part convector, part radiator. Nice!

Steam baseboard

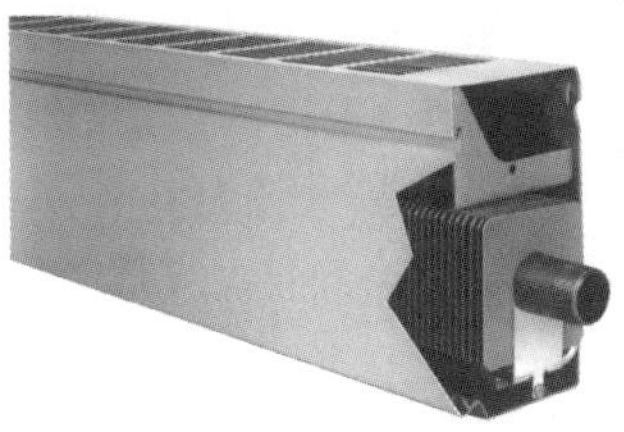

Courtesy of Slant/Fin

Let's say you don't like the look of any of the radiators we've talked about so far. You're remodeling and you'd like to have something that looks more modern. You're thinking about baseboard convectors, like the ones you see on hot water heating systems, but you're not sure if you can use these with steam.

You can! But there are some rules that you'll have to follow, and really think twice before you make this one a weekend project. Here's why.

First, the type of baseboard we use on steam systems is different from what we use on hot water systems. Hot water baseboard is copper; steam baseboard is either steel or cast iron, and the heater elements are wider on the inside because the steam and condensate have to share that space. Don't let anyone sell you copper baseboard for steam heating. The racket it will make will keep you up nights.

Next, If you have two-pipe steam, there will be a supply valve on one end and a steam trap on the other end of the existing radiator. If you switch to baseboard, the element will have to pitch at least one-inch in 20 feet downward from supply valve to steam trap (one or both of which will have to be relocated).

If you have one-pipe steam, however, things are not that simple. Steam condenses very quickly inside baseboard, and it's rare that you can get enough pitch to make it work as a normal, one-pipe counterflow system (where steam goes one way and condensate goes the other way). The steam will always be trying to shove the water backwards toward the air vent at the end of the baseboard, and I'll bet you the steam wins every time. I've been in houses where the water, propelled by the steam, hit the air vent so hard that it blew the vent right off the radiator. Pretty impressive force! In one very memorable house I visited, the water was splashing off the dining room ceiling. No kidding.

But don't give up; you can have one-pipe steam baseboard. You just can't do it as a standard one-pipe installation. What you'll have to do is pitch the element *downward* toward the air vent (one-inch in 20 feet) and drain the end that's opposite the supply valve back into a wet return (that's the line below the boiler's waterline). You'll also need to vent the end of the convector element. What you're doing is creating a hybrid one-pipe/two-pipe radiator (which is why the vent's allowed in this case). The steam enters through the supply valve, pushing air ahead of itself and toward the air vent, and then the condensate continues that way rather than returning the opposite way. This gets you around the problem of counter-flow and water hammer. Here's a sketch. In this case, I'm bringing the condensate back through a crawlspace and into a wet return for its journey back to the boiler.

You'll find both steel and cast iron baseboard at Slant/Fin (www.slantfin.com).

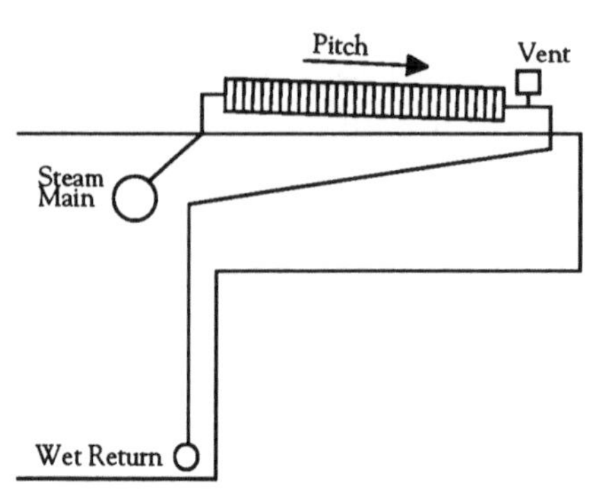

Okay, one more piping trick for you. If you're not able to run that wet return line all the way back to the boiler it's possible to go right back into the steam main with the radiator's return line. To keep steam from working its way backwards though the return line you'll have to use a water seal. This involves dropping the return line down about five feet and then bringing it back up into the bottom of the steam main. Like this.

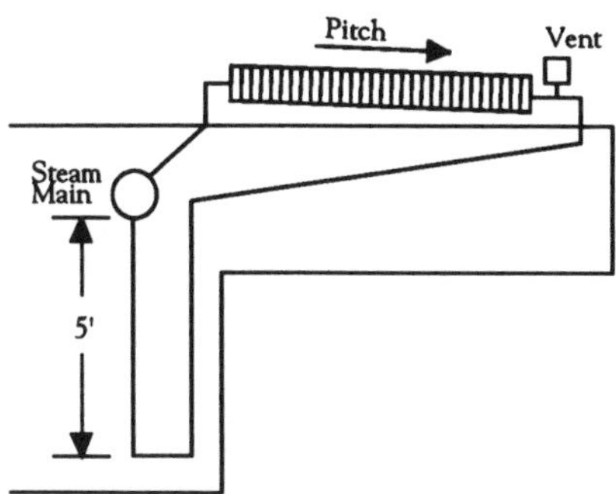

This is definitely a job for a pro because of the special tools you need to do it properly. He'll have to cut and thread the large main pipe in place.

If you're in a part of the country where there's not much steam, the pro may not know about this Dead Man's trick. Show him this drawing and tell him that this is what you want. He'll respect you when you show this to him, and you'll have baseboard heat that doesn't sound like the Fourth of July.

If he has any questions, send him to HeatingHelp.com for answers.

European-style radiators

Courtesy of Runtal

Up until recently, your choices in steam radiation has been pretty much vanilla, chocolate or strawberry, but there's now a line of European-style radiators that are stunning to look at and they work beautifully with steam. Go online to www.steamradiators.com and you'll see them in color. Great stuff!

Indirect radiators

EXCELSIOR PATTERN RADIATOR USED IN THE INDIRECT METHOD OF WARMING
American Radiator Company

METHOD OF INDIRECT WARMING AND VENTILATION, SHOWING ROTARY CIRCULATION OF AIR
American Radiator Company

And then there are these wonderfully strange radiators that hang heavy from basement ceilings, hidden within dark ductwork. These are remnants of the early days of heating when people were afraid of the air inside their homes. The indirect radiator heats a mixture of air that's already in the house with fresh air from outside. The two air steams flow into the ductwork, pass along and through the hot cast iron sections, and rise gently upstairs, usually through floor grates on the first floor. The upper floors of homes that have indirect radiators often have freestanding cast iron radiators, although I've also been in homes where the indirect heaters take care of all the floors (lots of ductwork in those places).

The air that enters the house through the ductwork also has to leave the house, of course. You can't just keep adding air; the place would blow up like a balloon. In the old days, the air left through the edges of the windows, which didn't seal tightly, or it scooted out under the doorways or through any other crevice that it could find. It's interesting to watch what happens when homeowners replace those old windows and seal up the cracks to conserve energy. If the air can't get out, it also won't come in, and at that point, the system depends entirely on the air that's inside the house recirculating from the upstairs to the basement.

As you can imagine, bringing lots of outside air into a house can be pretty expensive nowadays, so a lot of the people who own homes with these systems (and these are almost always in the jumbo-size homes) seal up the fresh air inlets. At that point, they need either a louvered door to the basement, or a return-air duct to get the upstairs air down to the basement so that it can be heated again. Funny how a louvered basement door can solve a heating problem, isn't it?

A quick story for you. Years ago, I was hired to look at one of the homes owned by Doris Duke. This was in New Jersey, a magnificent, 2,700-acre estate that made my head swim. The main house was one of the largest I've ever been in and it had an indirect heating system such as the one I just described (this one ran on hot water, though). Miss Duke's complaint was that the air rising from the bronze floor grates on the main floor was too cold. I checked it with a digital thermometer and it was only about 75 degrees, much too cool to get the job done.

I went over all the mechanical equipment with the plumber (he lived there; Miss Duke had one of every trade you can think of living on this estate), and all looked fine. We walked to the duct that fed upstairs and I opened the access panel. It was like opening the door of a pizza oven. The indirect radiator was hot enough but the air simply wasn't moving. "I don't understand this," the plumber said. "Heat rises. The heat *should* be going upstairs, but it just won't move."

"Heat doesn't rise," I said. "Heat radiates. Warm air rises."

"Well, it won't rise here," he said.

"That's because there's no cold air to replace it. You can't move air without replacing it with more air. Where does this duct go?" I pointed to the left.

"I think it goes outside," the plumber said.

"Where does it come out?" I asked.

"I have no idea. I don't do outside. I do inside."

We went outside anyway.

After looking around these gorgeous grounds for about ten minutes, neither of us could find the spot where the ductwork left the building. So we went back to the basement and measured in from the corner of the house to the spot where the ductwork went into the outside wall. Then we went outside again and measured in from the same corner, but all I found was topsoil and some newly planted ivy.

I had a hunch, so I started to stomp around on the ivy, and, sure enough, there was this hollow noise coming from beneath the plants. The plumber went to get a shovel from the gardener and we started digging. When we got down about six inches we hit the big sheet of Plexiglas that the gardener had used to cover the ground-level iron grate that led to the passage that connected with the fresh-air duct. We removed the Plexiglas, went back inside and both of us smiled as we felt the hot air wafting up out of those bronze floor grates.

No concerns about the increased fuel bill in this place! Nope, not at all.

That was a very good day.

Direct-Indirect radiators

Here we have a freestanding cast iron radiator that sits above a fresh-air duct that leads to the outside. A mixture of room air (entering from under the radiator) and outside air (coming up from the duct) meet at the bottom of the radiator and rise together as the steam gives up its heat. Most of these radiators have had their outside connections sealed to save energy. The room may overheat when you do this because these radiators are usually oversized; they're designed to heat both indoor and outdoor air, and that takes some doing. If the room is too hot, a thermostatic radiator valve, or a radiator enclosure will help (more on radiator enclosures coming up soon!).

Used radiators

You can try a local junk yard but you never know what you're getting there. I keep a list of used radiator dealers on our HeatingHelp.com website. These folks buy and sell all over the US and Canada and they test each radiator to make sure that it's sound and will work well without leaking.

Enter this on your browser and follow the links:

http://www.heatinghelp.com/newsletter.cfm?Id=120

And if you happen across another good source, let me know and I'll add it to the list. Thanks!

Refinishing radiators

The best (but not the cheapest) way to refinish an old radiator is to have it removed, sandblasted and then powder coated. The powder that they use in this process is a mixture of finely ground particles of pigment and resin. They spray this onto the radiator's surface and the charged powder particles stick to the electrically grounded surfaces. The radiator then goes into a curing oven where the powder fuses into a smooth coating. The result is absolutely gorgeous. You get a very even, and very durable, finish. To learn more, contact the Powder Coating Institute at www.powdercoating.org.

And if you're going to remove your radiators, be aware that some of them are about as heavy as a Mosler safe, so unless you're into living dangerously, get a pro to do the dirty work. They have these electric hand trucks that climb stairs, and they know how to use them. Worth every penny.

A lower-cost refinishing option involves lots of elbow grease. Get yourself some wire brushes in all shapes and sizes and have at it. And please take care not to breathe what comes off those old radiators. I'm guessing that one or more of those layers of paint has lead in it. Please wear a surgical mask while doing this work, and keep the kids away.

And once you're done scraping and cussing, you can use any good paint (latex or oil-base) to finish the job. Heat isn't an issue here since the surface temperature of the radiator won't go above 220° F., and that's well within the range of what a quality paint can take.

Does the color of a radiator matter?

Surprisingly, yes! But it's only the color of the final coat of paint that matters. The layers of paint won't affect the radiator's ability to move heat into the room because paint isn't much of an insulator. If it were, you could expect to save lots of money on fuel every time you painted a room in your home. Nope, it's only the color of the final coat that matters. And that's why you'll see so many steam radiators that are painted either silver or gold. Those two colors indicate that someone used a metallic paint. That's a special paint that contains tiny flecks of metal (either aluminum or bronze). When those flecks are on the hot surface of a radiator they affect the radiator's ability to radiate. This has nothing to do with convection (the movement of air across the hot surface). It has only to do with the radiant side. A freestanding cast iron radiator gives up about 60 percent of its heat by convection and the remainder by radiation. Radiation is responsible for the sensation you feel when you stand in front of a radiator (or a campfire) and the warmth flows over you like honey. You feel it on the front of your body, but not on your back. That's radiant heat. It's the same stuff that comes to us from the sun. It travels through the air without heating it, and will only warm solid objects.

The flecks of metal in the paint on the radiator's surface diffuse that radiant energy, and by quite a bit. An aluminum bronze color reduces the radiator's ability to radiate by 20 percent. A gold bronze paint will reduce it by 19 percent. Isn't that amazing?

And why would anyone want to do this? Well, remember earlier when I told you about the Spanish Influenza pandemic of 1918 and how, after that horrible winter, heating engineers began to size radiators to be able to heat homes with the windows open. As the flu panic faded, and the Great Depression arrived, people began to look for ways to save fuel. They closed their windows, and that's when they came to realize how ridiculously oversized their radiators were. So they used the paint with the metal flecks to tone down the radiator's ability to radiate. It was an inexpensive way to put a muzzle on the big beast.

And they also used radiator enclosures. Read on.

Radiator enclosures

These come in all shapes and sizes and styles. I've seen them made of wood, metal and even marble. Some of the marble ones are delightful because they often don't have enough holes through which the air can escape. The homeowner spends a small fortune and winds up being miserably cold. But, gosh, they look great!

The dimensions of any enclosure, and where the holes go, are both very important. Here are some drawings that show the effect on a radiator's ability to warm the air that passes over it, based on the size of the cabinet and the location of the air inlet and outlet. As you look at these drawings, think of the radiator as being uncovered. In that case, its output would be 100%. When you see a note reading, "Add 20%" that means that to get the same output from the enclosed radiator that you would get from an uncovered radiator of the same size, you would have to make the enclosed radiator 20% larger. That's easy to do when you're designing the system and putting this stuff in from scratch, but that's not what

RADIATOR ENCLOSURES

To enclose or partly enclose a radiator reduces its heat output and changes the distribution of heated air in the room. The additional surface usually added to column or tube radiation for various enclosures is indicated below.

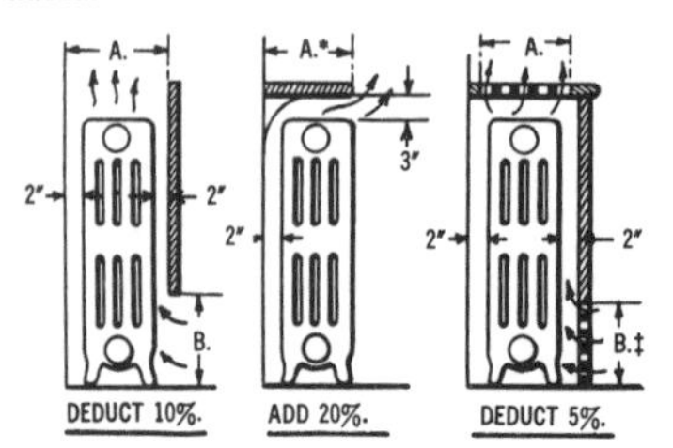

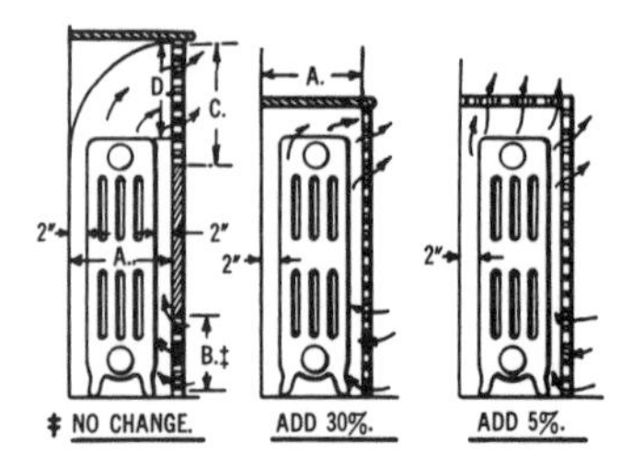

*If A is 50% of width of radiator, add 10%; if 150%, add 35%.

‡B = 80% of A. C = 150% of A. D = A.

Example: A room requires 50 sq. ft. radiation radiator recessed flush with wall, —50 ¢ + 20% = 60 ¢ radiator required. If radiator for same room is to have grille over entire face only, —50 ¢ + 30% = 65 ¢ required.

you're doing, is it? So read it this way. If you used that particular enclosure, you'd have 20% less heat in that room. Will that make you uncomfortable? It depends on how oversized the radiator is for the room in the first place.

Should a steam radiator get hot all the way across?

Steam radiators, like all radiators, are supposed to keep you toasty warm on the coldest day of the year. Where I live on Long Island, the day that heating engineers have in mind as the coldest (or what they call the **design day**) is the day when it's 10 degrees above zero and the wind is whipping around at 15 miles per hour. On that very cold day, that radiator will be just the right size. And that's the day when it's supposed to get hot all the way across.

But how about the days when it's not so frigid? Should that old steam radiator get hot all the way across then? Common sense tells you that if it did, you'd probably be much too hot. You'd be gasping for air and holding your stomach and twisting the supply valve closed. You might even be opening the windows and heating the neighborhood. So much for energy conservation, eh?

On a mild day, that steam radiator should only get partially hot, and that's perfectly normal. The thermostat will have shut off the burner before the radiators had a chance to overheat. When the steam stops moving, the radiator stops heating. It's as simple as that. If only half the radiator is hot and you're comfortable, that's the way it should be. Don't be concerned. A larger portion of the radiation is bound to get hot as the weather gets colder because the thermostat will keep the burner running longer.

Hot water radiators are different. They have hot water flowing through them whenever a thermostat calls for heat. They always get hot all the way across, every time. Not so with steam, so don't be concerned. It's normal.

The Pipes, The Pipes are Calling!

If you look at the pipes in your old steam system you'll notice that most of them are made of black steel. This was the standard in the old days. As I mentioned earlier, some contractors like to pipe replacement steam boilers in copper rather than steel nowadays because they can solder copper joints. They have to thread steel pipe together. Threading involves more work than soldering so it's more expensive. When a contractor quotes you a price for any steam work (a new boiler or any piping within the system), ask him what type of pipe he plans on using. I think steel pipe is much better than copper for steam heating work, but like I said, it *does* cost more to do it this way. I think it's well worth it, though, and here's why.

Copper expands *much* more than steel, and that's an important consideration, seeing as there's so much twisting and turning going on in these systems. That's what those odd pipe angles are all about. They allow the pipe to expand and contract and twist without upsetting the pitch needed for proper condensate drainage. If the pitch isn't right, you'll wind up with noisy water hammer and uneven heat (and high fuel bills).

When a contractor uses copper tubing instead of threaded steel pipe for those special angles, and then solders the joints together, the twisting from expansion and contraction will often cause the pipes to come apart over time. The problem generally shows up shortly after the guaranty period runs out.

But you did get a good price.

And this problem with expansion and contraction is especially critical within the near-boiler piping. The contractor that quotes your job based on using copper for the near-boiler piping is doing himself a favor, but he's certainly not helping you. I've seen too many of these jobs come apart over the years to believe otherwise.

Another problem with copper is that it leaches out and

into the boiler as time goes by. This can cause dielectric corrosion within the boiler, and that shortens the boiler's life. You wind up buying another boiler.

But you did get a good price.

My feeling? Pay a bit more and have it done the right way. After all, how many times in your life will you be replacing your steam boiler?

True story (it happened in Northern New Jersey). A contractor gets a call from a woman who wants him to quote on replacing her steam boiler. He measures all the radiation and looks over the entire job. He asks her about any problems she's had with the system and he talks to her about her comfort needs. Then he gives her a price and she tells him she'll get right back to him, which she doesn't do.

A few weeks go by and the contractor figures that she must have gotten someone else to do the job. But then she calls, and she's in a panic. She pleads with him to come over right away. It's an emergency!

So he drives to the job and races up the front walk. She's waiting for him by the door and she hurries him in. The house is filled with steam. He's confused. She leads him to the basement steps. "Hurry!" she says. "I didn't know who else to call. The other guy refused to come back. He said he was too busy."

The contractor gets to the boiler and finds the source of the steam. It's spewing from the joints of the PVC plastic pipes that the other guy used for the new boiler's near-boiler piping (you can't make this stuff up). The plastic is hanging from the boiler like limp linguine and the contractor asks the woman why she let the other guy use plastic pipe on a steam boiler. She tells him that the other guy was a thousand bucks cheaper than he was. "Lady," the contractor says, "didn't you ever stop to wonder *why*?"

But she never thought about that. She just wanted a good price.

Which she got.

Always ask about the type of pipe the guy's planning on using. Always.

Here's another way to test your prospective contractor. Once he tells you he's going to use only black steel pipe for your job, look him in the eye and ask him if he's going to use cast iron or malleable fittings. If he's a good steam man he'll probably chuckle and tell you that he's going to use cast iron fittings, and then he'll explain the difference.

If you hold a hammer behind a cast iron fitting and then hit the other side of the fitting with another hammer, the cast iron fitting will crack and you'll be able to get it off the pipe (everything in a steam system rusts solid after awhile). If you try this trick with a malleable fitting, though, the heavy hammer will bounce off the fitting and hit you in the forehead, providing a lesson you need learn only once.

Which means that cast iron fittings are better for steam work. They break when hit, and that's exactly what a pro will do when he's trying to remove that old boiler. It's literally impossible to unscrew steam pipes from steam fittings after a few years unless you can crack them. They fuse together into a huge, rusty macromolecule. This, then, is the advantage of cast iron fittings. They break.

If the contractor insists on malleable fittings, know that this really isn't a problem for you. It will, however, be a problem for the next contractor who tries to remove that boiler someday when it has grown old and earned its retirement.

Another thing. Steam pipes need to pitch so that the condensate can drain when the system shuts off. Without the proper pitch you'll probably get a water hammer racket when the system first starts. Here's the rule for pitch:

If the steam and condensate are heading the same way, the pipe has to pitch one inch for every 20 feet of travel.

If the steam and condensate are heading in opposite directions, the pipe has to pitch one inch for every 10 feet of travel.

If the steam and condensate are going in the same direction there will be a drain at the end of the main pipe. If there's no drain, then things are going in opposite directions. We call this **counterflow**, and as you can see, the required pitch doubles. Most of the Dead Men did their best to avoid this counterflow situation because there's more of a chance for error when the steam is opposing the condensate in a counterflow situation. This is where the expression, "Go with the flow" comes from. It does.

If you have the asbestos removed from your steam pipes there's a good chance that the pitch won't be right once the workers are done. These folks are great at what they do, but they're not steam experts. They may not reset the pipe hangers properly if they remove them to get at the asbestos. That can leave you with water hammer once the weather gets cold. And this situation is aggravated if you don't replace the asbestos with fiberglass insulation. The combination of bad pitch and cold pipes is a good recipe for water hammer, especially on start-up.

And please don't hang things on your steam pipes. Heavy objects can cause the pipes to sag. I once visited a house that was plagued with water hammer. When I went downstairs I found a heavy bag (the kind boxers use) hanging from the steam main.The bag was filled with sand and weighed more than George Foreman.

And don't let your kids do chin-ups on the steam pipes because that can also make the pipes sag (they'll only attempt this during the summer, though).

The angle and the pitch that the short pieces of pipes between the steam main in your basement and the risers that go up to your radiators is also critical. If you think the pipes are in your way, and you'd like to raise them a foot or so, it might be time for you to consider another type of heating system. Any contractor that tells you he can raise those pipes to get them out of your way is a knucklehead. He'll get paid and you'll have no heat next winter. Trust me on this.

The return pipes that run along your basement floor, or those that are buried beneath your basement floor are filled with water all the time. That water is acidic and all the goop that's in the system will settle into these low points. And then there's that acidic fly ash in the concrete. Put all of that together and you can see why "wet" returns are always a cause for concern.

My approach to these lower-system pipes goes like this:

1. If a pipe is below the boiler's waterline, assume it's clogged.

2. If a pipe runs beneath the floor, assume it's leaking.

3. These things are guilty until proven innocent.

And most of them *are* guilty.

Controls

The Basic Controls

One of the really nice things about a steam heating system is that the controls aren't complicated. They watch over temperatures (both air and water), and they also look at the pressure inside your boiler as well as the water level. That's it.

Some of these things you should leave alone, and some of them you should touch every week. Let's take a look at each one.

The Thermostat

A homeowner called to say that he was having problems with his steam heating system. I asked him what type of system he had. Was it one-pipe steam or two-pipe steam, or maybe vapor? He said, "Hang on a minute. I'll go see." He came back a moment later and said, "It's a Honeywell system."

That was the name on his thermostat. Honeywell.

Get it?

This has happened to me more than once and it's perfectly understandable. Most homeowners think of the thermostat when they think of their heating system. You chilly? Turn up the thermostat! Too hot? Turn it down!

The thermostat's a simple device. Most use a bimetallic coil to sense the air temperature in the room. A bimetal is two different metals that are fused together. These metals expand at different rates so when the air temperature changes, the coil either tightens or loosens and trips a switch that sends a signal that starts the burner.

Steam systems work best when they have basic on-off thermostats, such as the one you see in the photo. Thermostats that automatically turn the temperature down at certain times of the day or night (commonly called **night setback thermostats**) sometimes cause problems with steam systems because they allow the pipes to get too cold too often. Keep in mind that the original system probably ran on coal, a fuel that burned nearly all the time. The Dead Men who installed your system never intended for the pipes to go completely cold for large parts of the day or night. When you switch to a setback thermostat you may find that the pipes hammer and knock all of a sudden. This is because the cold pipes are making a lot more condensate, and that condensate isn't getting out of the way fast enough.

If you combine a setback thermostat with a system that has had the asbestos removed from the pipes and not replaced with fiberglass insulation, then you're going to hear the anvil chorus every time the thermostat clicks back on.

Can you use a setback thermostat? Yes, but first have a knowledgeable pro look over the whole system. That's the best way to avoid problems. And let the pro do the wiring. Thermostats can be tricky.

Thermostats also have that **anticipator** that I mentioned earlier. The anticipator anticipates when the room air temperature is near, but not at, the temperature at which you've set your thermostat. It then stops the burner before it can bring the room to its set temperature. There's a flywheel effect with all heating systems and if you don't stop them prematurely they'll override the setting. The anticipator does that. It's comparable to stepping on the brake *before* you get to the stop sign, rather than trying to go from 40 miles per hour to zero in a tenth of a second. Do that and you'll skid through the intersection, right? Anticipators are cautious drivers. They keep you from skidding past your desired room temperature.

The right anticipator setting for your system depends on the amperage that's running through that particular circuit, and this will vary from system to system. The only right way to set the anticipator is with a meter, and that's something that the pro should do for you.

The Pressuretrol

This is the wonderful device that senses the steam pressure and starts and stops the burner. It's the first thing your knucklehead will go for when he visits your basement.

The pressuretrol works with the thermostat and the two make a fine team. It's like Mr. Upstairs and Mr. Downstairs. The thermostat (Mr. Upstairs) senses that your house is getting chilly and sends a signal to the burner and make some steam.

The water boils and the steam heads out into the system, pushing air ahead of itself and out of the air vents. If everything is working as it should, the system will soon fill with steam and your radiators will get hot. The burner continues to run and more steam enters the system. When this happens, the pressure of the steam will build because that's the way gases work. If you fill the system with a gas and then add more gas you'll get pressure. The more steam you add, the more pressure you'll get. It's like putting air in your car's tires (only a *lot* hotter).

Okay, enter the pressuretrol (Mr. Downstairs). He's set up to sense two pressures. We call the first pressure **cut-in** and the second pressure **cut-out**. Your boiler will operate between these two points as long as the thermostat is still calling for heat upstairs.

On the most common pressuretrol, you can see the cut-in pressure by looking at the slide on the front of the pressuretrol.

The scale reads in pounds per square inch, and for just about every house in America, the setting should be bottomed-out at .5 psi. That's one-half pound per square inch. Crank it down.

Now the other setting (the cut-out) is not apparent. You have to open the cover of the pressuretrol to see it, and you have to do a small bit of math.

See that round plastic wheel? It's marked DIFF, which is short for DIFFERENTIAL. If you take the cut-in setting (in this case, that's .5 psi) and add the Differential (let's say we set the wheel at 1 psi), then your steam system will operate between 1/2 psi and 1-1/2 psi. In other words, the cut-in pressure plus the differential pressure equals the cut-out pressure. Some pros will call this an **additive pressuretrol** because you add the pressures to get the cut-out pressure.

Just about any steam system in any home will operate best at a cut-in setting of .5 psi and a differential of 1 psi. Crank it down, and if your system doesn't heat well at these settings, it's probably because the air isn't getting out of the pipes. Look for those big main vents near the ends of your steam mains. If you don't have them, get them.

There's another type of pressuretrol that you may have, although it's not as common as the one I just showed you. This one is a **subtractive pressuretrol** and you'll know it because it has two scales on the front instead of that plastic differential wheel on the inside.

One of the scales is labeled Main and the other is labeled Differential. The Main scale indicates the **cut-out** setting for your boiler. To get the **cut-in** pressure, you have to subtract the Differential setting. For instance, let's say the Main scale is set at 2 psi and the Differential scale is set at 1-1/2 psi. Now, subtract 1-1/2 from 2. That leaves you with 1/2, right? That's the cut-in pressure, and it means that your boiler will now operate between 1/2 psi and 2 psi.

Always try to run your system at the lowest pressure possible. That's where you'll get the best performance.

The Vaporstat

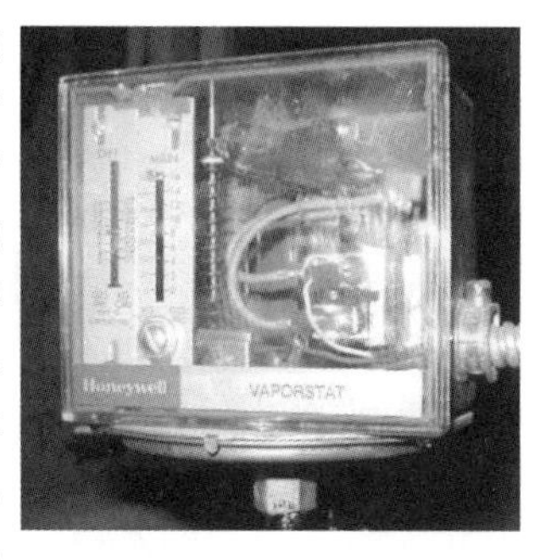

A vaporstat does the same thing as a pressuretrol, but it does it far more accurately, and in a lower range of pressure. And here I'm talking *ounces* of pressure rather than pounds of pressure. While more expensive, a vaporstat is the control that I think works best on residential systems, especially those two-pipe vapor systems.

Make sure you're venting the system very quickly (from big main vents) if you're using a vaporstat, though. If you're not, the burner might short-cycle.

(A word about pigtails)

This little gizmo is appropriately named, don't you think? A pigtail is a curved piece of pipe that goes between a boiler and a control or a gauge. Its job is to fill with water and to keep the latent heat of the steam from reaching through to the control (or gauge). The fewer the Btus inside the control or gauge, the longer that piece of equipment will stick around.

Some controls contain mercury switches. That's a glass tube containing a few drops of mercury. When the tube tilts on its bimetallic element, the mercury slides over and lands between two electrical contacts, completing the circuit and starting the burner (with the help of the relay). It's important that any control with a mercury switch (such as a vaporstat) be level so that the mercury tips and makes the electrical connection only when it's supposed to do so. When a pigtail gets hot it will expand and try to straighten itself out, so make sure that the pigtail on your system is facing in the right direction. When you're looking at the front of the control you shouldn't be able to see the "hole in the donut." If you can see the hole, turn the pigtail 90-degrees so that you can't.

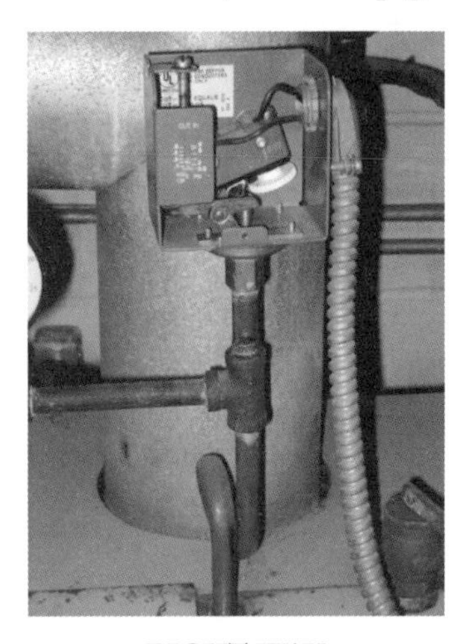

RIGHT WAY

WRONG WAY

The Aquastat

An aquastat senses the water temperature in the boiler and starts and stops the burner as needed. Your boiler may or may not have an aquastat. If you're using your boiler solely to heat your home, there's no need for an aquastat; the boiler will operate off of the thermostat and the pressuretrol. However, if your boiler is also making your domestic hot water through a tankless coil (which is a coil of copper pipes that's submerged below the boiler's waterline) then it will have an aquastat. The aquastat will keep the boiler water hot (but not so hot that it will make steam) so that you can bathe and do the dishes during the summer when you don't need any steam in your radiators.

Hands in pockets, please.

The Relief Valve

The maximum safe pressure for your steam boiler is 15 pounds per square inch. The relief valve watches over that pressure like a home plate umpire and blows its top if the pressure should get to that highpoint. This is a very good thing because if your home-heating boiler is running at that high a pressure something is definitely out of control. The relief valve releases the boiler's pressure and that can save your life because it might prevent a boiler explosion.

There should be a pipe coming from the relief valve that extends downward to a point just above the floor. That pipe directs any escaping steam to a place that's safer than right in your face. *Never* attach anything to the end of that pipe that might stop the steam. And *never* run that pipe outdoors where a block of ice might form in the pipe during the winter and do the same.

Ideally, your relief valve will serve your boiler for years and never pop. It's like a police officer's weapon. Good to have, and even better if it's never needed. It should be tested each year, and replaced immediately if it doesn't open properly, and shut tightly again after opening.

Some contractors are afraid to test the relief valve because if it doesn't seat properly after they test it (and that often happens because of the dirt in these systems), they think you'll blame them for "breaking" it and demand a new one at no charge. Believe me, if that relief valve doesn't seat properly after the test, it's the one thing you *don't* want on your steam boiler.

Ask the contractor to test it, and tell him that you realize it may not shut tightly after the test. And if it doesn't, have him install a new one. Then pay him, and thank him. Your family is now safer.

The Gauge Glass (not a control, but pretty important!)

This lets you see the water level inside your boiler. It's normal for the water to bounce about three-quarters of an inch or so when the boiler is making steam. If the bouncing is crazier than that the boiler probably needs attention. A lot of things can cause bouncing and none of them are good. Have a pro look at it.

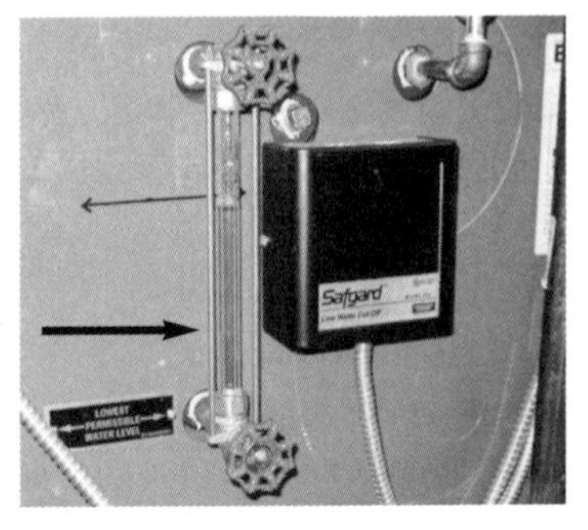

If the water appears brown, don't be too concerned. Steam systems are open to the atmosphere and they're constantly rusting. That's what you're seeing. If the gauge glass gets so dirty that you can't see the water, have the glass replaced. These are available at most plumbing and heating supply houses. If you want to do this yourself you'll probably have to cut it to fit between the metal fittings, so get yourself a round glasscutter as well.

The Low-water cutoff

Before we had automatic burners that fired gas or oil there was only coal and wood and those were scary fuels because if the water level in the boiler fell too low there was nothing there to stop the fire. This was part of what led to the boiler disasters of the 19th Century. In the early 1920s, though, companies came up with both automatic burners and electric low-water cutoffs, and these made the world a safer place.

This control is very important because it senses the water level in your boiler and stops the burner if that level gets too low. Low-water cutoffs are required by code everywhere and they're there for your protection. A dry-firing boiler (one that has flame in it, but no water) can get hot enough to start a fire in your home. There's also a risk of explosion, should someone add water to your boiler while it's dry-firing.

A very popular type of low-water cutoff is the one in the picture above. McDonnell & Miller (www.mcdonnellmiller.com) makes these and there are millions of them out in the world. There's a metal float inside that cast iron chamber. The float is connected to an electrical switch that's wired in series with the burner. Flip that switch and the burner will stop. The iron float chamber attaches to your boiler with those two pipes that hook-up with the gauge glass. The water level that's in the chamber will match fairly closely the water level that's inside your boiler. Should the boiler lose water, the float should sense the loss and flip the switch.

But since this is a steam system, and because steam systems are constantly corroding, crud is going to find its way into that float chamber, which is like a quiet lagoon compared to the storm that's going on inside the boiler. It's your job to clear that crud out of the float chamber, and it's a job you must do once a week during the heating season. You'll do this with the **blow-down valve** that's piped to the bottom of the float chamber. Place a metal bucket under the blow-off valve and flush about a quart of water from the unit (plastic buckets can get mushy and fall apart so make sure you use a metal bucket). Do the blow-down while the boiler is running to make sure that when you flush, the float actually does drop and shut off the burner. And please be careful when you do this. That water is *hot*!

This is another type of low-water cutoff. It's called a

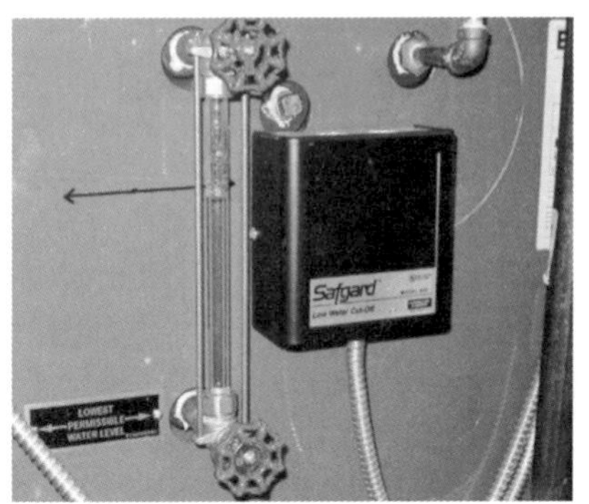

probe-type cutoff and it's electronic. It screws into the side of your boiler in a tapping that the boiler manufacturer provides especially for this type of control. That's important because the probe has to be at just the right level inside the boiler to do its job.

The probe sits in the water and a tiny electrical charge passes from the tip of the probe, through the water, and back to the metal surface of the boiler. The water is actually completing the circuit, just as the mercury did in the mercury switch. If water should drop off the probe for more than a few seconds, the circuit will break and the burner will shut off. And please don't be concerned about mixing electricity and water here. It's perfectly safe. The current is very low voltage.

There's no need to blow-down a probe-type low-water cutoff, but someone does have to remove the device from the boiler once a year (during the warmer months) and clean the probe. Leave yourself a note and make sure you have this done; it's very important.

When it comes to steam, I'm a belt-and-suspenders kind of guy. And although it's not required by code, I think that every steam boiler should have two low-water cutoffs, preferably a float-type and a probe-type. They should be in different places on your boiler. This doubles your protection, and since low-water cutoffs are *much* cheaper than boilers, it makes sense. They can also save your life. Besides, the cost of that second low-water cutoff disappears in the price of the job when you're replacing your old boiler. Please consider it.

The Automatic Water Feeder

The automatic water feeder acts as a back-up to your low-water cutoff. Since all steam systems are open to the atmosphere they're going to lose water to evaporation, and to minor leaks. The automatic water feeder watches the minimum waterline and replaces the water lost to evaporation and minor leaks. It does this by using a mechanical float that trips a switch and opens an electric solenoid valve. The instant the valve opens, water begins to flow. It brings water to the proper level and then shuts off.

Courtesy of McDonnell & Miller

The feeder maintains a safe minimum water level inside your boiler, but if you have one of these on your boiler, please don't think that *everything* is automatic; you still need to check your boiler once a week. This type of feeder maintains a *minimum* water level, not an operating water level. You have to set the operating level by hand when the system is shut down and cool. The operating level (when the boiler is off and cool) should be about two-thirds up in the gauge glass. You set this by opening the manual feed valve or by pressing the bypass button on the feeder. When the boiler is making steam, the level will drop to about the one-third-full point in the gauge glass. This type of automatic water feeder will look for a point that's a bit lower than that, very close to the bottom of the gauge glass. That's the minimum.

Many steam systems don't have automatic water feeders. In this case, the homeowner fills the boiler by hand when it needs it. The biggest danger here is usually the telephone. It tends to ring while you're feeding the boiler. You run upstairs to answer, chat for a while, then remember that you're supposed to pick up the kids at school. When you return there's water pouring out of the air vents on those second-floor radiators, which, of course, causes you to say a few disparaging words such as, Ah, shucks! And so on.

The Electronic Water Feeder

Courtesy of Hydrolevel

I like this type of electronic water feeder because it has features you don't find in most electric or mechanical feeders. This one, for instance is Hydrolevel's VXT (www.hydrolevel.com). It keeps track of how many gallons of water enter your boiler. If you keep a notepad near the feeder and take a look at it once a week you'll be able to tell if the boiler is suddenly taking on too much water, which would indicate a leak in the system (probably from a buried pipe). This feeder also has the ability to feed past the minimum waterline to the operating water level. It will also wait before feeding, should the low-water cutoff shut off the burner. This gives the condensate a chance to work its way back to the boiler and helps prevent overfilling. Smart stuff!

The Feeder/Cutoff combination

This unit, also by McDonnell & Miller, does two jobs in one. It's an electrical low-water cutoff and a mechanical water feeder. The mechanism inside the feeder is similar to the float valve in a toilet tank. The float's buoyancy keeps it on top of the water. As the water level drops, so does the float, and when it does, it opens a mechanical valve that allows water into the boiler.

Notice the blow-down valve at the bottom of the large float chamber. Make sure you open that valve and flush about a quart of water once a week. If you don't do this, the chamber will fill with goop that will become as solid as a sidewalk, and that's guaranteed to hold that float up when it should be down. The float is also the thing that controls the electrical cutoff switch. *Please* don't neglect that weekly blow-down.

This type of mechanical feeder is sometimes prone to problems when the feed line clogs. That happens because there are minerals in the cold feed water that come out of solution as they approach the boiler in the feed pipe (and all of these feeders should always use cold, not hot, water). This is especially true in hard water areas. As the minerals build up, a form of heating arteriosclerosis sets in. The inside diameter of the feed pipe narrows and this causes backpressure as the feed water flows through the line. It's sort of like trying to blow through a straw while you're pinching it closed. The backpressure can often be severe enough to keep the mechanical float from shutting tightly. This, of course, causes the boiler (and sometimes the whole system) to flood. A flooded boiler produces wet steam, and if the water rises into your pipes and radiators while you're away on a month-long vacation to Tahiti, you will not need a heating contractor upon your return; you will need Jaques Cousteau.

If you do go away regularly during the winter, there are high-water alarms that can call your heating contractor (usually through your alarm company), or flash a light on the outside of your home that your nosey neighbor will notice. Check with your contractor about these.

Thermostatic Radiator Valves

Thermostatic radiator valves for steam systems come in two parts. One part is a spring-loaded, normally open valve that's either going to allow, or not allow, the flow of steam or air. The other part is the control head, which snaps onto the valve. The control head contains a chemical that's very sensitive to changes in air temperature. Together, the two parts make up the thermostatic radiator valve, or what we call a **TRV**, for short.

TRVs don't need electricity to work; they're completely self-contained. Here's how they get the job done.

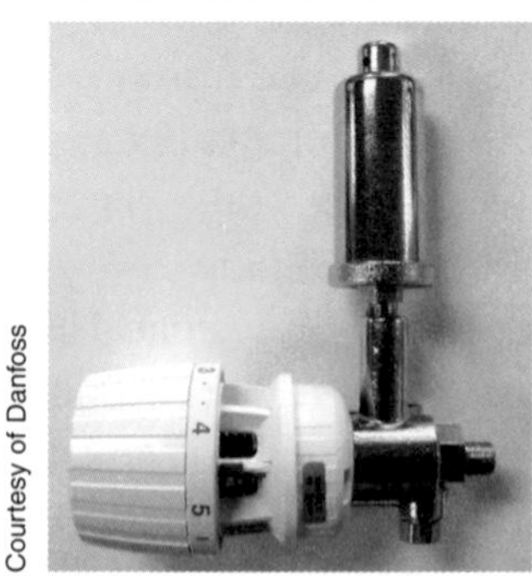
Courtesy of Danfoss

When you have one-pipe steam, the supply valve at the radiator has to be either fully open or fully closed because the steam and the condensate share that space. Because of this, we can't use a TRV to control the actual steam; we have to come at it from another direction. We place the normally open, spring-loaded valve (which is pretty small) between the radiator and the radiator's air vent. The control head snaps onto the valve and constantly monitors the temperature of the air in the room. There's a dial on that control head that you can turn to select whatever temperature you'd like in the room, usually between 50 and 90 degrees F. (the dial may just read Warmer/Cooler, however, or 1 through 5).

The steam enters the radiator at the call of the electric thermostat, which should be in the coldest room in your house. And never use a TRV in the same room as the electric thermostat. That room becomes the control room for the entire system. The boiler will run until the electric thermostat is satisfied, and the TRVs will keep the rest of your place from overheating. That's their job. If one side of your house is sunnier than the other, the TRV on that radiator will sense the warmth of the sun streaming through the windows and keep the radiator from overheating that area. If you're roasting a turkey, the nearby TRV will sense the heat from your oven and turn down that radiator.

If you have a radiator that's cold all the time, however, a TRV won't make it any hotter. They're there strictly to keep things from overheating. If the room is cold, it's probably a balance problem that's caused by air not being able to escape from the pipes.

Okay, so the steam enters the radiator, pushing air ahead of itself and out the radiator's air vent. The air has to go through the normally open, spring-loaded TRV valve to get to the air vent, of course. As the room air temperature comes up to the setting you've chosen for the TRV's control head, the chemical inside the control head will flash into a vapor and expand (just like a steam trap's element, or the float inside an air vent). The pressure of the expanding vapor will push down on a metal bellows (which looks like a little accordion), and the bellows will push the stem of the spring-loaded valve until the valve closes. If no more air can escape the vent, no more steam can enter the radiator. It's as simple as that.

Now, there's one challenge with a one-pipe-steam TRV and that's vacuum. Once the valve shuts and no more air can escape from the vent, the steam that condenses in the radiator will form a vacuum. That vacuum can pull more steam into the radiator, causing the room to overheat. To get around this, most TRV manufacturers build a tiny vacuum breaker into the body of their valve. A vacuum breaker does just that – it breaks vacuum by allowing air back into the radiator. It's like a one-way door that's hinged inward, toward the radiator. During normal operation, the steam pushes the door closed, but when a vacuum forms, the door gets sucked inward and air goes barging right in. It's a small detail, but it keeps the radiator from overheating. If you buy these valves for your one-pipe radiators, make sure you ask for the ones with the built-in vacuum breakers.

Installing TRVs on one-pipe radiators is a job a handy person can tackle. Just read and follow the instructions, and don't use one in the room where the electric thermostat is. Ideally, that thermostat should be in the coldest room. If it's not, have a licensed electrician move it for you.

If you have two-pipe steam, you can use this type of TRV.

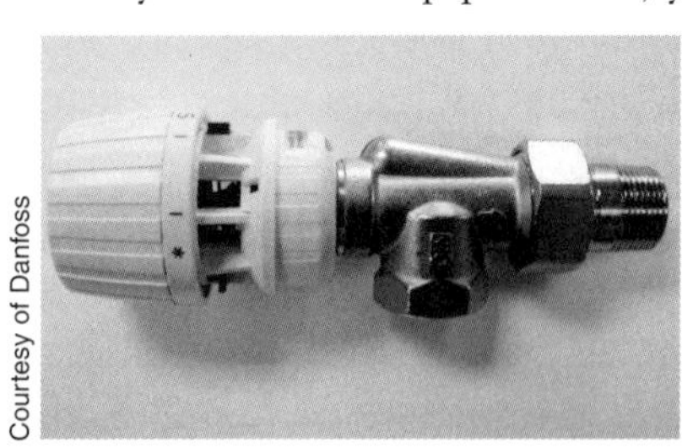

Courtesy of Danfoss

Here, the steam and the condensate are separated by the two pipes so you can use a TRV in place of the supply valve. It will control the amount of steam that enters the radiator, based on the air temperature in the room, and the setting that you've selected for the TRV.

There are no vacuum breakers built into this type of valve because we're controlling the inlet side of the radiator. Once the valve is shut, no more steam will enter. They work the same way as the one-pipe variety, except there's steam, not air, moving through them.

TRVs also come with built-in and remote sensors. If your radiator is inside an enclosure, or if you have convectors (those are the cabinet-type heaters), make sure you use only the TRVs that have remote sensors. The device needs to feel the air on the outside of the enclosure, not on the inside.

Unless you have the right tools and lots of experience with old piping systems, I would not tackle a two-pipe-steam TRV retrofit. The valves and the old pipes usually don't line up perfectly and it takes experience to make it all come back together properly. This is definitely a job for the pros.

Combined with proper air venting, TRVs can make a big difference when it comes to comfort. They're well worth looking into.

Things you can do

Here's a list of things that a handy person (that you?) can do for himself or herself

Flush the low-water cutoff: If you have a low-water cutoff with a float (see above) make sure you flush it once a week during the heating season. Do this by opening the valve at the bottom of the float chamber, and do it while the burner is running. You only need to flush about a quart of water from the thing so don't overdue it. You want to make sure that any sediment that might interfere with the float doesn't accumulate inside the float chamber. If you're diligent about this, flushing a quart a week during the heating season, you'll add years to the life of your boiler by protecting it from dry-firing (flame in the boiler, but not enough water). You might also save your life.

And here's a tip for you. When you're done flushing, close the valve quickly and watch the gauge glass. The water, which will leave the glass during the flushing, should bounce right back into the glass when you shut that valve. It should come up quickly, which indicates that the lower pipe that connects the low-water cutoff to the boiler is free of debris. If the water trickles back into the gauge glass, call a pro and have that pipe replaced. It's dangerous.

Lower the steam pressure: The pressure on most residential steam boilers is much too high and that wastes fuel and can cause system problems. Look for your pressuretrol (there are pictures of the most popular ones further back in the book). Remove the cover and look inside. You'll probably see a small plastic dial that's marked DIFF (for Differential). Turn this to its lowest setting. Now, look on top of the pressuretrol where you'll find a screw. Turn this and watch the slide gauge on the front of the unit. The screw moves the slide up and down. Turn the screw so that the slide goes down. Bottom it out and you'll be at the right cut-in and cut-out pressures for your house.

And if your house won't heat at that pressure, it's not because of the pressuretrol; there's something else wrong with the system – probably trapped air from missing or failed air vents, or the lack of insulation on the pipes.

If you have a pressuretrol that reads, "Differential is Subtractive" then set the MAIN scale indicator to 1-1/2 psi and the DIFFERENTIAL scale to 1 psi. That will allow your system to operate between 1/2 psi and 2 psi, which is the most any house needs.

Insulate the pipes: Tuck them in and the steam will go further. I've talked a lot about this already and I hope I've convinced you of the importance of pipe insulation. Steam is a gas that wants to be a liquid. Don't help it by having naked pipes. If you do, you'll have less heat where it counts – upstairs.

If you have asbestos on your steam pipes now, and it's in good shape, you can wrap it with plastic. If you're thinking about having it removed, please don't do this yourself. Get a qualified asbestos abatement company to do the work for you. Your future health is worth far more that what you'll pay those folks.

If the pipes are bare, cover them with insulation. Use fiberglass insulation, not the rubber type that you'll find in home centers. There are companies that make shaped pipe insulation that opens like a clamshell. It comes in various sizes and you'll have more luck finding this at a good plumbing-and-heating supply house than you will at a home center. You'll have to know the size of the pipe when you go there so measure the outside diameters of all the bare pipes that you're going to insulate. The supply house will also have insulation for the elbows and tees throughout your system. If you're not concerned with appearance, you don't have to use these. You can cover the corner elbows and the tees with plain insulation (like you'd use inside your walls) and duct tape it all together. Not pretty, but it works.

For that matter, if you don't care how it looks, you can cover all the pipes with plain insulation and duct tape. Your choice, but make sure they're tucked-in.

Keep in mind, though, that in some buildings the pipe that goes from floor to floor just might be the radiator. This is especially true in apartment buildings where the only heat in the bathroom and the kitchen comes from that bare vertical riser. Don't insulate these pipes unless you feel they're overheating the rooms.

Check, and if necessary, change your air vents: Got radiator or convector vents? You can handle these. They just screw into the radiator. Clockwise tightens them; counter-clockwise loosens them. Use a pliers, or your hand and do this only when the steam is off. *Please.*

Look the vent over and shake out any water that may be stuck inside of it. If it doesn't gross you out, blow into the threaded part of the vent to see if the air comes out the other end. If it seems to be leaving very *slowly*, try cleaning the vent. You'll do this by placing the vent in a pot of vinegar and bringing it to a boil on your stove. Hold your nose. The vinegar is a mild acid and it does a nice job of breaking down deposits that usually wind up in air vents.

Put a bit of Teflon tape around the threads of the vent and screw it back into the radiator. Make sure the vent hole points toward the ceiling. Keep the steam pressure *low* (that's the pressuretrol's job). See what happens.

If you're still having the same problem, it's time to buy some new air vents. You'll find the best ones at plumbing-and-heating supply houses, and they're not all the same. As with most things in life, you get what you pay for.

And the same goes for main vents. These can be pretty expensive. You can change them yourself if you have a big wrench. The same things that apply to radiator vents apply to main vents, but on a heftier scale. Chances are that what you're taking out is from a company that's out of business so take the vent to a good plumbing-and-heating supply house, show it to the counterman and ask him for a suitable replacement.

If you get stuck on any of this, stop by our website at www.HeatingHelp.com. We don't sell vents, but if you post your question on the Wall (our bulletin board) and show us a photo, we should be able to tell you exactly what you're dealing with, and what would be a good replacement for it.

Pitch a one-pipe steam radiator: If need be, you can do this, but be careful. Watch your back and don't do it alone. Lift (or lever) the end opposite the supply valve and have your helper slip a piece of wood, metal or checkers under each of the two radiator's feet (on the free end). If you're older and wiser, you'll be the one sliding the checkers. Let the young guys do the heavy lifting.

Repack the supply valves: If the stem around your radiator supply valve is leaking you can probably repack it. Go to the plumbing-and-heating supply house or to your local hardware store and get some graphite packing. Remove the valve's handle and unscrew the packing nut. That's the one on the valve stem. Wrap the graphite around the stem and screw the packing nut back down. Put it all back together and see how you did. Good luck!

And make sure the steam's off when you're doing this. *Please.*

If you have two-pipe steam and your radiator valve is

leaking you may not be able to repack it. Many of those old systems ran into vacuum so that the homeowners could get the maximum amount of Btus out of their coal fires. Water boils at a lower temperature in a vacuum, and to keep the vacuum from breaking, the Dead Men used what they called **packless valves**. These had diaphragms to seal the valve stem. Once this type of valve starts to leak, it's time to replace the whole thing, which is a job that's best left to a pro because the new valve might not rough-in the same way as the old valve did, and it's pretty easy to snap those old pipes if you don't have the right tools and experience.

And one last thing you can do: Know when to keep your hands in your pockets. A steam system is like a child's mobile. When you touch one part, everything else starts swaying. If you're not sure what will happen when you touch something, don't touch it. Get help.

Things you probably shouldn't attempt

Repairing radiators: I'm putting this in this section mostly because it usually involves a lot of work and it will probably wind up breaking your heart in the end anyway.

Here's why.

I told you earlier about the difference between the screwed nipples and push nipples that connect the sections of old radiators. Recall? The oldest (and usually the prettiest) radiators had the screwed nipples with those left-hand/right-hand threads that pulled the sections together as the Dead Man turned the nipple with a special wrench. If you have those types of push nipples, trust me; they're not available *anywhere*.

If you have push nipples, you're in better shape (and so is your radiator). You can still get these. And you'll know you have push nipples between those sections by the long tie rods that run through the radiator sections. If you see the tie rods, it's possible to replace a busted radiator section, but only if you're feeling real adventurous.

First you have to pry the thing apart. Take a long, loving look at that ancient radiator. Feeling lucky?

Let's say you are. Once you get it apart (without cracking it), you'll have to work the push nipple out of the section in which it's still stuck and then call Oneida County Boiler Works in Upstate New York (Phone: 315-732-7914). They'll want you to send them the old nipple (regardless of its condition) and they'll take good care of you. They regularly help people all across the country, and they assure me that Oneida is the only company around that still supplies these fittings. "If people could get 'em any closer to home, they wouldn't be calling us!" one of their guys told me. I believe him.

Once you get the new nipple you can lube it up and then try to get the whole thing back together again. This requires patience, a gentle touch, brute strength and a certain amount of Zen. And maybe some alcohol (for yourself). Work those tie rods and hope for the best.

Got a small leak? Try J-B Weld, which you can find in most home centers and hardware stores. But to fix the leak, you'll first have to be able to get at it and prepare the surface. Before using the J-B Weld, you'll have to fully drain the radiator and remove any paint, primer, or rust from its surface. Next, you'll have to thoroughly clean the surface with a non-petroleum-based cleaner, such as acetone or lacquer thinner, removing all dirt, grease and oil. Then you'll have to rough up the surface with a file, mix the two elements of the product in 50/50 proportions, and apply it to a thickness of no less than 1/32". Don't get any on your skin or in your eyes. Finally, let it dry for at least 15 hours, and see what you've got.

Can you do that?

The challenge, of course, is that an antique radiator can have more nooks and crannies than a Thomas' English muffin, and a good leak knows where to hide.But if you're in love with that old radiator, it's certainly worth a try.

Give it a lot of thought, though. Old radiator repairs can break your heart.

Moving radiators: I don't think you should do this because I've heard from so many homeowners who have tried. They speak of pain, and in some cases, of a potential new Winter Olympics event – Radiator Luging. This is where you ride a five-hundred pound radiator down a flight of stairs and through the living room wall. Go ahead; make a memory!

Professional heating contractors know how to disconnect old radiators without breaking the pipes. Those fittings are often rusted together and need to be heated (without burning down your house) or otherwise coaxed into letting go. Then there's the massive weight to consider. Contractors use expensive, motorized hand trucks that climb stairs. It's safer this way. Hire someone to do this work for you.

Move Piping: Don't do it. Period. Those pipes in your basement that keep hitting you on the noggin are at that height for a reason. Steam pipes have to pitch so that the condensate can roll back to the boiler. They're also a certain size and you should never alter this. If you do, the pressure drop across the system will change, the steam and the condensate will go to war, and you'll wind up listening to the steam-heating equivalent of artillery fire. It's called water hammer.

The risers that come up through your floors and connect to radiators can't be angled off this way and that just because you'd like to put the new couch where the old radiator is. The steam and the condensate don't care about your decorating ideas. Move those pipes and the steam and the condensate will get even with you.

If you hire someone to work in your house and they tell you it's not a problem to raise pipes or reroute them around your living room, and that they can do the whole job in copper and raise everything up a foot or so, know that you are having a conversation with a world-class knucklehead.

Change steam traps: You might not want to tackle this either because so many steam traps are obsolete. Parts are available for most of these, but they come from companies that make kits. You have to know what you're looking at to get the right kits. If you guess wrong, you're out a bunch of money and you've still got the problem. Besides, pipes break. If you don't have the right touch to know when you've put as much torque on an old steam trap as it can take you can be in deep trouble very quickly. A pro will feel it, back off, and try another approach. An amateur will snap a pipe every time.

Pros know how to get these old traps apart, and how to either fix them, or replace them when they can't be fixed. They also know that a bad steam trap can pass steam into a return line, making it appear that working traps are defective. When it comes to fixing traps, it pays to leave it to the pros. It really does.

Convert the entire system to hot water heat: I'll tell you what's involved with this in a little while. If you're using the old pipes and radiators it can be quite challenging.

Replace the boiler: The same goes for the boiler. Steam boilers need to be piped with steel pipe, not copper (or PVC. *Please.*). You need special equipment to cut and thread steel pipe. The near-boiler piping has to be perfect if the boiler is to work to its potential because, nowadays, near-boiler piping is a part of the boiler. The system needs to be *thoroughly* cleaned afterwards. It's tough work, and depending on where you live, it may even be illegal to replace your own boiler.

Beyond that, there's the safety of your family to consider. If the burner isn't set up properly, if the connection to the chimney, and the chimney itself, isn't just right, you could all die from carbon monoxide poisoning.

This is not a good place to save money. Please leave it to the pros.

Replace broken pipes: *Why* did that pipe break? Was it acid corrosion? Water hammer? A wacky thing called erosion-corrosion (caused by an over-active condensate pump)? Is the broken pipe buried under your cement floor? Got a jack hammer? Where will you start looking for the leak? Got an infrared heat detector?

Pipes break for many reasons, and when they do, someone has to figure out why it happened so that, hopefully, it won't happen again. And that's a job for a professional heating contractor. Worth every penny, those guys. Trust me.

Anything else that makes you stare at the ceiling at night: When in doubt, get help. Stop by our site at www.HeatingHelp.com. Listen in to the chat on the Wall, our very active bulletin board. Ask questions. If you're going to get in over your head, we'll tell you.

And you'll be hearing from pros from all over the country who have nothing to gain or lose from your job. Visit with us and ask questions; we'll help.

And if you're looking for a steam-heating pro, you'll find a lot of good ones hanging out there.

Clean Systems Are Happy Systems

Steam systems get dirty because they're constantly corroding. Rust happens. It sinks to the bottom and also gets stuck in your air vents, steam traps, strainers, nooks, crannies and anywhere else it can find. Sediment from fresh feed water also settles to the bottom, and the minerals in the water come out of solution and cling to the system's hotter surfaces, just as they will in a coffee maker (you clean those with vinegar, don't you? Vinegar's a mild acid).

Since everything settles to the bottom of the system, there will come a time when the wet returns (the pipes below the boiler's waterline) will need to be cleaned. This is not a job for the fainthearted, and this can lead to quite a mess if you don't know what you're doing. Beyond that, the pipes might start falling apart.

To give you a sense of what's involved in cleaning wet returns, I asked ten professionals who regularly visit the HeatingHelp.com website how they handle this tough job. The question was:

When cleaning out the wet returns on a steam system, how do you keep from making a mess on the customer's floor?

And here's what they had to say.

1. We use newspaper, red rosin paper and kitty litter. We are as careful as possible, but spills do happen. Because we know this, we also bring along a heavy-duty mop and wringer bucket. This, along with a good floor cleaner, enables us to leave the work area cleaner than we found it. Most of our clients are pleasantly surprised by the effort we put forth to keep their floors clean. It makes for repeat business!

2. If it's not already there, I'll pipe in a boiler drain and ball valve on each leg of the wet return. This allows me to flush the line and direct the mess through a garden hose to a safe place outside the building. The cost of the few fittings is worth it. It shows customers that we care about their system and keeping their home clean. On new boiler installations, I install the boiler drains and valves as part of the installation.
3. I would suggest a couple of ideas. Fit an adaptor to connect a pump discharge hose and run it to a safe location. Maybe hire your local drain cleaner with a water jetter and big sucker truck.
4. A lot of speedy dry, a good vinyl tarp and a wet vac do the trick. Even better, change those old drain valves to full-port ball valves and give it a good flush.
5. We use a big shop vacuum to suck out the water as we

flush the return out. Just make sure you have a safe place to empty the vac because that will make a mess too.

6. What we did with one system in an old school (built in 1911) was to use a sewer rooter service. The man inserts the cutter of the correct diameter and cuts out the crud. As he pulls the cable back, we wipe the cable with rags, and as the glop is pulled out we pull it into a ring of rags, weaved loosely together to make a well. Then we suck up the mess with a wet/dry vacuum by introducing some water as we go to make the crud into a slurry.

   If you have a really big job, as we did some time ago, think about hiring a sewage service company. The vacuum is so strong that the operator can hold the nozzle at the pipe opening and suck the crap up before it can even leave the area. Also, because the truck suction is designed for semi-solids, it will take just about anything that is wet and can create a vacuum at the nozzle. It does not do a good job on dry material, however, and you have to wet down any friable material if you want the hose to take it.

   One person I know used the ring off the diamond drill (the ring that takes away the water as you cut). He used a vacuum on it, and it worked well as long as it was made into a fairly watery mess. However you do it, though, it is a lousy, dirty, crummy job – just the sort we all like to do, especially if you have a mouthy helper!

7. We try to talk them into new piping below the waterline, which costs about as much as a thorough cleaning will. Barring that, we'll either use heavy plastic to capture the boiler "ink" or a shop vac to keep the mess contained. New piping is the better method in most ancient replacement cases, and a cloth rag or duct tape stuffed in, or taped over, the old opening helps avoid the old age dribbles.Better yet, we invite their kids to the basement for finger painting on the plastered recreation room walls (just kidding!).

8. I've come across this many times. I started by draining the boiler (I'm assuming that there is no drain cock on the return line, right?). Next, I'd break the 'L' at the Hartford Loop, put in the suction hose from a small pump, and suck out whatever I can. Now the options:

Option 1: If there is enough play to pick up the return at that end, I do so. I then place a brick under it to give it some back pitch. I crack the fitting, remove it and install a tee with a drain cock, and then I reconnect everything and continue with the job.

Option 2: If there's no play in the return (such as with an underground pipe), this can be a real pain. I'll chop the concrete or cut the floor just enough to expose the fitting and remove the debris all around the riser coming up from the floor, and then I'll remove that riser by cutting a notch at the threads, remove it, collapse the threads in the fitting, unscrew the riser and install a shorter riser with a tee and a drain cock.

Option 3: If the return is horizontal, but has no play (and it won't if it passes through a partition), and there is room, I'll crack the tee and use the riser that goes to the Hartford Loop as a lever. I'll pull down, making that vertical piece horizontal, and use gradually smaller (in height) receivers such as roasting pans to get the water out. So you would pull down, catch some water until the pan fills, then quickly push it back up, dump the water and then repeat as necessary. Usually a few old towels at the very end of the process are all that you'll need.

Now naturally, you have to inform the customer that whenever you have to crack a fitting it might damage another part of the return, but if that damage does occur, the return needed to be replaced anyway. Now, when I talk about "cracking the fitting" I'm saying to hold back against the force of the striking hammer on the opposite side of the fitting with a heavier hammer. You know, for 1-1/2" pipe, and smaller, a 28-ounce hammer with a three-pound back-up works well. If the return is larger, then hold back with a sledge. The idea is to only create a crack in the cast iron fitting, not to bust out a piece of it. Just enough to produce a hairline crack, and then it will not offer any resistance. Hope this helps.

9. If we need to avoid any spillage at all, we cut a piece of 6-mil heavy plastic to fit in the area. We place it so it goes up any walls to a few inches in height and spread it out over a more-than-adequate distance, based on the most water we estimate might spill or splash. We always put down plenty of plastic to play it safe. We buy the 6-mil plastic in 20' wide x 50' or 100' long rolls and cut off whatever we need. We can sometimes use the plastic more than once if it hasn't been cut in a bad spot for the next job. Some small cuts can be duct-taped to be damp-proof. Then we cut a line into the plastic to where any penetrations come out of the floor and make some crisscross cuts there, being careful to make the outside diameter of these cuts a little smaller than the penetration so it fits tightly and we tape it in place. Duct tape works well for this. Then we put down some absorbent rags around where we will open the pipe. We always keep a plentiful supply of rags in the shop and on every truck. We bring extra when we know we're going to do a job that requires them. And before we open the pipe, we make sure we have more than enough catch buckets ready to handle the flow, and we plan where we're going to dump the buckets, and we make sure we have a clear and protected path to that spot. And we always do this with a two-man crew so that we have an extra pair of hands ready to keep the job as neat as possible, and to help swap out the full buckets for empty ones. Also, we don't fill the buckets all the way to the top to prevent spillage. Finally, after we're done, we dry any spillage and our boots before we start walking around. We clean up, and, 'voila', a satisfied customer is usually the result.

10. At the union or elbow, break apart the pipe and using the elbow and nipple, raise the pipe up so it can drain into a bucket. Use a shallow pan and pump if the return is very low, or simply replace the return with new pipe and save yourself the headache.

That's what's involved. They each have their own way of getting the job done, and, as you can see, it often involves specialized equipment. And that's why I think you should leave this job to them. It doesn't have to be done that often. If you're hearing banging in the pipes that wasn't there last year, or if you suddenly see water squirting from air vents, especially the vents in your basement near the ends of your steam mains, you probably need to have the wet return cleaned out or replaced.

And if you're having your boiler replaced, ask the contractor to install a shutoff and a drain valve in the main wet return line, just before it connects to the Hartford Loop. These two valves in the return will give the installer, as well as future service people, a way to flush your return lines with much less mess. It's well worth the little bit of extra expense. A good steam contractor will probably do this without being asked.

Now, here's another concern. New boilers contain a lot of oil and that oil will rise to the surface of the water when the boiler is making steam. This will lessen the quality of the steam that the boiler sends to your radiators. It will cause your fuel bills to be higher than they should be, and it might create water hammer if the steam gets wet enough. Reputable boiler manufacturers include instructions for cleaning and skimming with every boiler they sell. When you're making your deal with the contractor, ask to see these instructions, and ask if he will do the cleaning the way the manufacturer has laid it out.

If the contractor tells you that this isn't necessary, and that he can just add a can of chemicals to the boiler, know that he is a knucklehead. Chemicals cannot make 100 years worth of goop turn into Evian water. That stuff must be thoroughly flushed from the system and the oil in the boiler must be skimmed from the surface of the water.

Skimming involves using a special tapping at the boiler's waterline. The contractor will install a few fittings in that tapping, which he'll use to skim the oil from the surface of the water. It's like spooning the grease off of gravy or spaghetti sauce. It takes nearly a full day to do this job properly, and a good contractor will ask to be paid for the time he spends working for you. It's worth it.

Chemicals for steam boilers do a few things. Some have an oxygen-scavenger that removes the oxygen from fresh feed water. This helps avoid oxygen corrosion (the process that eats holes in your boiler at the waterline). But I'm thinking that if the boiler is taking on enough feed water to need chemical protection against oxygen in the water then there must be a leak in the system. Right?

So wouldn't it make more sense to find and fix the leak rather than just keep throwing chemicals at the problem?

Other chemicals cling to goop that's floating on the surface and cause it to sink to the bottom of the boiler. The goop then needs to be flushed from the bottom of the boiler. Leprechauns will not show up in the middle of the night with little wheelbarrows to carry it away. Someone else has to do it. And don't let the contractor tell you that it's okay for the goop to stay in your boiler. That's like doing your laundry without a rinse cycle. Hey, you paid for a new boiler. Why have it filled with old goop?

And a good steam contractor will always install a valve at the very bottom of your boiler, in the **mud leg**. The mud leg is where the mud winds up. The boiler comes with a plug in the end of the mud leg. After a few years, that plug fuses with the boiler and even Archimedes couldn't get it out. Get a valve in there from the git-go and your boiler will last much longer.

For cleaning, most manufacturers recommend a strong soap called trisodium phosphate, or TSP for short. TSP is a great boiler cleaner and it's available at most paint stores. You use one pound per 50 gallons of boiler water. It has to be mixed in hot water and injected into the boiler. You run the boiler for hours and then thoroughly (and I mean *thoroughly*) flush it from the system. If you don't get it all out the boiler water will foam like a Holiday Inn hot tub and you'll be miserable. Please don't try this on your own.

If TSP isn't available in your area (because of the phosphate content), a good steam pro might use a commercial soap called MEX, which is phosphate-free and just as good.

Ask about these things when you're talking to prospective contractors. The ones in the know will respect you for asking, and you'll scare the heck out of the knuckleheads.

Vinegar isn't a cleaner, but some pros may use it to lower the pH of the boiler water to prevent foaming. If you have one-pipe steam, the odor of the vinegar will wind up upstairs as the system vents its air. An alternative to vinegar is lemon juice or orange juice. They all work well at lowering pH.

How often does a steam boiler need to be cleaned? It depends. Certainly when it's first installed. And then when there are problems that weren't there last year.

How to take a hot water zone from your steam boiler

Okay, you decide to remodel your old-house. Maybe you're planning on a new kitchen, or you're going to enclose that back porch and turn it into a year-round sitting room. You're going to have to heat that new space, and maybe you're thinking that all you'll have to do is add another steam radiator or two.

So you get in touch with a local heating contractor and he stops by to take a look. He sees your old steam system and then glances at you with an expression that you, as an old-house owner, know well. It lies somewhere between Hopelessness and Great Expense. You gulp.

"Can you do it?" you ask.

The contractor's face shifts to somewhere between Pity and Even Greater Expense. "Not so easy," he says.

Part of the challenge you're both facing here is that the existing steam pipes might not be able to handle the additional load. You also have to see if you can tie into that main in the basement with the proper size pipe and still get the correct pitch to the new radiators. And then you have to wonder if the new radiators will be compatible with the old system piping. And what effect will all of this have on the old boiler? Will the new piping bang and spit water when you're done? And if so, what then?

All in all, it's not so easy, but let's face it, what you're looking for is to heat a new space, and there's really no reason why you have to tie into the old steam piping to get that new zone. In fact, it's possible to add a brand-new *hot water* zone to that old steam system and even put it on its own thermostat.

Depending on where you live, however, you may run into pros that don't know that this can be done, so I'll tell you about it, and you can tell them about it, and if anyone has any questions, you can get more advice at www.HeatingHelp.com. Good deal?

Okay, here's how a pro can add a hot water zone to your steam heating system. First, remember that your steam boiler is like a teakettle. It's partially filled with water and it uses the space above its waterline to make steam, which then races off into the piping in search of a way out (that being the air vents on the radiators, and the vents near the ends of the main piping). To add a hot water zone to that old steam system, the pro will have to grab hot water from the boiler at a point below the boiler's waterline. He'll use a circulating pump to move the water between the boiler and your new hot water radiators, and he'll return the water to another tapping below the boiler's waterline. That return tapping must be in a spot on the boiler that's not too close to where the supply tapping is. This ensures good circulation across the boiler. If the supply and return tappings are too close together the water will just scoot through your boiler and not be in there long enough to pick up the heat it will need to satisfy your new zone. And this is where the tools come in. Securing those tappings in a boiler, particularly an older boiler, can be a real challenge. This is *not* a weekend project.

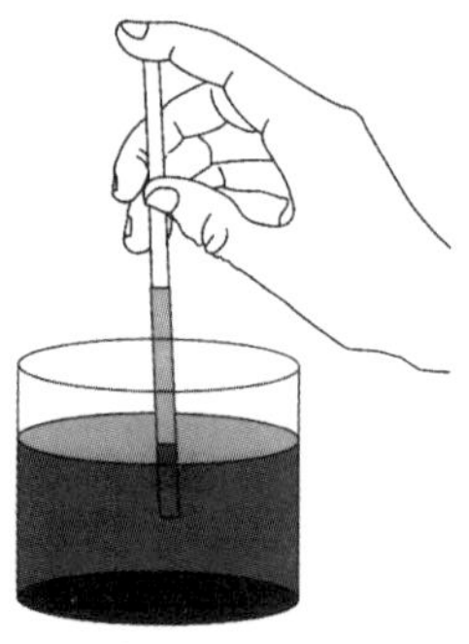

Now I know you're probably wondering how the water is going to stay in the new zone's piping if that piping is higher than the boiler. To answer this mystery, you'll need to get yourself a glass of water and a drinking straw. Got it? Good! I'm now going to ask you to do something that you've probably been doing since you were a kid. Put the straw in the glass of water and then place your finger over the top of the straw and lift it from the glass.

The water stays in the straw, right? How come? Because the weight of the air (the atmospheric pressure) pushing the water up into the straw is greater than the static weight of the vertical column of water that's trying to fall out of the straw. Take your finger off the top of the straw and the air will suddenly have access to both ends of the water column. Gravity will take over and the water will fall out of the straw. The Principle of the Straw will allow you to put a hot water zone on even the *second* floor of your house. Even if you live in Denver.

But the Principle of the Straw is also why you can't have any air vents in your new piping and radiators. If air gets in, the water will fall from the pipe and wind up back in the boiler (and you can't use the circulating pump to fill the zone each time; they don't have enough power to do that). The piping to and from the new radiators should ideally be a continuous loop so that the pro can fill it with water before starting up the zone. He'll do this with a purge system, which is nothing more than two shutoff valves and two hose bibbs –

one of each on the supply and the return, and below the boiler waterline. Like this.

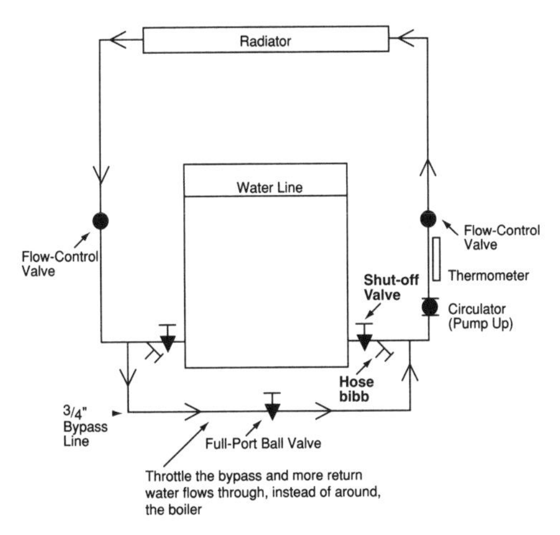

The shutoff valves will be between the boiler and the hose bibbs. He'll close both shutoff valves and open the hose bibbs. Then he'll put a hose on one of the bibbs and fill the piping and radiators with water. When water flows from the other bibb, he'll know that he's done. He'll shut the bibbs and open the shutoff valves. The water will stay in your new zone, just as it stays in the drinking straw. Pretty cool, eh?

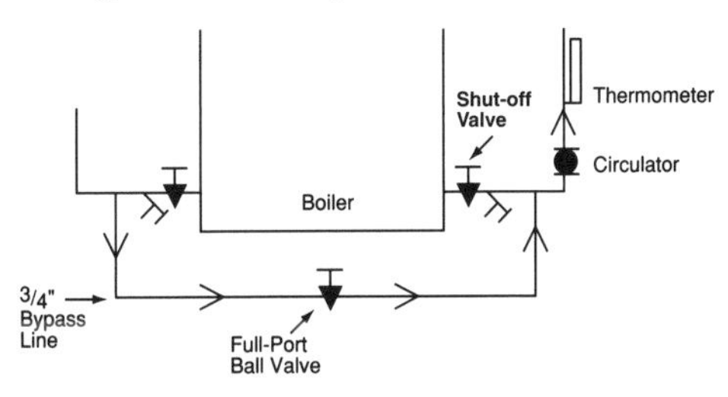

Now here's the piping trick that makes it all work. When the boiler is making steam, the temperature of the water inside the boiler is going to be hotter than 212° F. The circulator will pump this hot water out of the boiler and up into the zone. As it does this, the circulator will be adding some pressure to the water, so the water that's at the top of the system will remain liquid. But when the circulator shuts off (and that happens when the thermostat in your new zone is satisfied), the scalding-hot water suddenly loses that pressure and can flash into steam at the high point of the system. When water flashes to steam, it increases in volume nearly 1,700 times. This sudden expansion of steam can shove the water from the radiator and piping and dump it into the boiler. And accompanying this will be sounds you will long remember.

To keep this from happening, the pro will pipe a boiler-bypass line between the return line and supply line of your new hot water zone.

The bypass will allow some of the water that's returning from the radiator to go around the boiler and join with the hot water that's leaving the boiler. The result will be water that's about 180° F. (when the boiler is making steam). This piping technique mimics what goes on inside a kitchen sink's single-lever mixing valve. It blends hot and cold water to deliver a mix that's just right – not too hot and not too cold. And in your new zone, that blending ensures that the water at the top of the new zone can't flash to steam when the circulator shuts off.

To control all of this, the installer will use three devices. A **thermostat** in the space will sense the air temperature and start the circulator on a call for heat. The water will move past an **aquastat**, which is like a thermostat, but a bit different in that it senses water, not air, temperature. If the water temperature is 180° F., or hotter, the burner will not fire. You'll recall we talked about aquastats earlier. You don't normally see them on steam boilers unless the boiler also has a tankless coil for making domestic hot water, but we'll use one here for this hybrid application.

The third device is called a **switching relay**. Its job is to start the burner if the aquastat senses a temperature below 180° F., and to stop the burner before the boiler can make steam. This allows your new zone to operate independently of the steam system. You won't have to heat your entire house (with the steam system) when you just want to warm the new space with the hot water zone. And your steam system will still work off its own thermostat as it always has.

The only other thing we need to add, and these will go on both the supply and return lines, are **flow-control valves**. These are weighted (or spring-loaded) check valves that will keep the hot water in the steam boiler from rising (by natural convection) into your new zone when your new thermostat is not calling for heat.

Finally, I have to tell you about a few limitations with these hybrid systems. First, there's only so much heat you can take from your boiler before you won't be able to make steam. You can safely use a third of the total output of the boiler (its Gross rating). Make sure the contractor does a heat loss calculation on the new space to see what it needs. It can't exceed one-third of the boiler's total Gross load. For instance, if your boiler has a Gross rating of, say, 100,000 Btuh, you could have a hot water zone that would use up to 33,000 Btuh. Don't guess at these numbers; be sure.

Another option available to you with this hybrid system is to use it to provide hot water to an indirect domestic hot water storage tank. An **indirect** (that's what we call it) is like a large Thermos that holds your domestic hot water. It gets its heat from your boiler, usually through a coil that's submerged in the indirect's tank. Following the same piping techniques, you can use the water in your steam boiler to heat the water in the indirect tank.

One thing I'll caution you against, though, is using this system to provide hot water for a radiant heat zone. These below-the-floor systems are becoming more and more popular, but the water in a steam boiler is too dirty to be flowing through the tight confines of the plastic or synthetic rubber tubing commonly used in radiant systems. If you want radiant, the pro can still set up a zone off of your steam boiler, but he'll have to use a heat exchanger and a second circulator. There are also more controls involved in this. And more money, of course.

Finally, pay a bit more to get a bronze-body or stainless-steel circulator. It will last longer than a standard iron-body circulator in the slightly acidic boiler water. It's your best value in the long run.

If you decide to go this way and you have more questions, visit us at www.HeatingHelp.com and just post your questions on the Wall.

Thinking About Converting Your System to Hot Water?

First, begin with the type of radiators you have. Do they have push nipples across both the top and bottom sections? If they do, there's a chance you can have your system converted to hot water. If they don't, you're out of luck (at least with those radiators). Steam is lighter than air so it will rise in an old-fashioned radiator, pushing the air down and toward the air vent. If you try to use those radiators on hot water, however, you'll find that air is definitely lighter than water. It will shoot up to the tops of the radiator sections and stay there. You'll have a tough time trying to fill those things with water. And unless you can do that, you won't have any heat.

Your option would be to have each radiator section drilled and tapped for an air vent. If you have a 20-section radiator, you'll have 20 air vents poking their little heads above your antique radiator. And what will your in-laws have to say about that?

If you have one-pipe steam, and if you have the tube-type radiators (the ones with the push nipples across both the bottom and the top), it's possible to convert the radiators to hot water usage, but you'll need a second pipe to serve as a return to the boiler from each radiator. This isn't as scary as it seems because, nowadays, we have this wonderful plastic piping called PEX (an acronym for cross-linked polyethylene) that your contractor can snake through tight spaces as if it were electrical wire. A nice option, should you choose to go this way?

You'll also have to remove the steam air vents from each radiator and plug the openings. Then you'll have to have hot water air vents installed in the top of each radiator to get rid of the air on startup. These vents will go in the big plug that's at the top of the final radiator section. You'll have to change the plug to accommodate an air vent, however.

Things are a bit easier if you have a two-pipe steam system. The supply and return pipes are already in place. The contractor will have to remove the guts from the steam traps so that the water can flow freely. If you have an old vapor steam system, he'll have to know what he's looking at and remove any orifice devices, tiny hidden metal balls, little check valves and other oddities that are sprinkled throughout those systems.

Which reminds me of an unhappy contractor who called here once. He had converted a steam system to hot water and told me that when the new hot water circulating pump ran the whole system whistled like a tea kettle coming to a boil. Right away, I thought that he had forgotten to remove the guts from the steam traps. These can narrow the passage so much that the water squeezing by makes a sound like the one you get when you let the air out of a pinched balloon. Imagine living with that.

"No," he said, "I didn't forget about the traps. I looked for them but there aren't any traps on any of the convectors. I'm positive about that."

"There must be something," I said. "There has never been a two-pipe steam system that didn't have some sort of device to keep the steam from reaching through into the return lines. What does it say on the convectors?"

"It says Trane," he said.

So I got out my old catalogs and looked up the Trane convector (these are the same folks who make the air conditioners nowadays, by the way. They've been out of the steam business for many years). I found what I was looking for and read a bit.

"You're in trouble," I said.

"Why?"

"There's a tiny orifice built into each of those convectors. That's what's making the whistling sound."

"Where is it?"

"Deep inside," I said.

"Can I get at them? Drill them out? How big a hole should I drill?"

"I don't know," I said. "Trane never intended for you to be doing what you're doing, and all of the people who invented this stuff are currently dead."

"Uh, oh," he said.

I'm not sure how it all worked out because he never called back. I think a good steam heating contractor would have spotted those Trane convectors right at the start, though, and asked the right questions before taking on the job. Good contractors do that; they look around, they know what they're looking at, and they ask a lot of questions – and not just of you, but also of their peers. On the HeatingHelp.com website, contractors are constantly asking each other for help in identifying the old and the weird. These are the good guys. They work first with their heads before working with their hands.

So once the radiator lines are clear of trap guts and/or orifices, the next thing that has to be removed from your two-pipe system is the wet return that comes back to the boiler from the end of the steam main. You read earlier about what a messy job that can be. This pipe has to go because there's no longer a need for it. The radiators each have their own return line, and these will be the lines that we use for the hot water conversion.

Once the pipes are all in place, the contractor will balance the flow rate across the system. This evens out the heat. He may set it up so that you can have zones throughout your house. This will be a more expensive job, but it also saves on fuel and provides greater comfort. There are also controls that will sense the outdoor air temperature and vary the water temperature in your new hot water system to match the heating needs from moment to moment. These controls work like cruise-control on a car, putting in exactly the amount of energy required at any time. Nice stuff.

But before any of this happens, the smart contractor will first test the piping and radiation to see if it can all take the higher pressure that's in a hot water system. That higher pressure comes from the weight of the water. The higher you stack it in the system, the more it weighs at the bottom. This is one of the reasons why so many old buildings in our cities are heated with steam. Take the Empire State Building, for example. It's heated with a few pounds of steam pressure, but if the Dead Men had installed a hot water system there back in 1929, that system would need to withstand about 540 pounds per square inch at the ground floor radiators. Better hope there are no leaks!

So your contractor will test the integrity of your steam system by raising the steam pressure to about 10 psi. Then he'll walk the whole system, looking for leaks. If there are any leaks, it's good that he finds them when there's steam in the pipes and not water. He'll repair or replace any pipes or radiators that need to be repaired or replaced and then he'll take it from there.

He'll also survey the entire system before giving you a price. He'll measure all your radiators and figure their output at the relatively cooler temperatures that you'll have with the hot water system. Steam systems run at 215° F. The typical hot water system runs at an average temperature of 170° F. Will you be able to heat your house with the lower temperature? The only way to know is to do a thorough heat loss calculation, and a complete radiator survey, which a good contractor will do before giving you a price. And depending on its condition, you may be able to use your old steam boiler. The contractor will have to remove all the steam equipment (gauge glass, pressuretrol, low-water cutoff, relief valve) and install hot water heating equipment, such as one or more circulating pumps, zone valves, a new relief valve, an expansion tank, and new controls.

There's a lot involved in this and that's why I think you're wise if you leave it to the professionals.

And here's how to find one of those.

How to find a good steam heating contractor

So, I've been telling you to find a pro and to avoid the knuckleheads. How will you tell one from the other. Here's a checklist of things to do as you negotiate with a heating contractor. Believe me; the good guys will thank you for being this thorough. An educated consumer is their best customer, especially when it comes to steam. Here goes.

✓ Walk the contractor through your home and discuss your comfort concerns. If it's too hot in this room and too cold in that one, let him know. If there are gurgling or banging noises in the pipes or radiators, mention this, and ask the contractor to suggest a remedy. Don't expect a new boiler to magically solve your system-related problems. Explain to the contractor that you're buying *comfort*, not just a new boiler, and that you expect the entire system - boiler, pipes and radiators - to work properly when the job is done. At this point, a good heating contractor will most likely explain your options. Don't be surprised if those options add to the cost of the job. Many steam systems, as you now well know, suffer from ailments that have nothing to do with the boiler. When the contractor suggests these changes, ask if he will guarantee the results. A good heating contractor will. If he's not willing to stand behind his work, there's a good chance he's a knucklehead. Call someone else.

✓ Don't let the contractor base the size of your new boiler on the size of your old boiler. He *must* measure your radiators and analyze their ability to condense steam. If he bases the size of your new boiler solely on the size of your present boiler, he's discounting the possibility that something may have changed during the past 70 or 80 (or more!) years. He's also acknowledging that the original installer was infallible (and we don't even know who that guy was). If the contractor says, "Well, the old one worked well for years so let's use the same size," know that this guy is not someone who pays attention to details. Show him to the door.

✓ Also, don't let the contractor base the size of your new boiler on a heat-loss calculation of your home. Heat loss calculations mean absolutely nothing when it comes time to replace an old steam boiler. These calculations mean a lot when you're sizing the steam radiators, but you're not replacing the radiators, are you? Remember that steam is a gas that will eagerly condense on cold metal. The boiler must be able to make enough steam to reach the furthest radiator before all the steam turns to water. In other words, the boiler's ability to produce steam *must* match the system's ability to condense steam. There's no way around this. If the new boiler is too small, parts of your home will probably always be cold and you'll burn more fuel than you should. If the boiler is too large, the burner will short-cycle, run inefficiently and the burner parts and controls will wear out long before their time. And again, you'll burn more fuel than you should. If the contractor doesn't take the time to survey and carefully measure your pipes and radiators, he doesn't understand steam heating. Tell the knucklehead to have a nice day and then show him the door.

✓ The piping around a modern steam boiler is crucial to the production of dry steam (steam that contains no more than two percent water). If the steam is wet, it will condense before it reaches all of your radiators. You'll wind up with high fuel bills and uncomfortable rooms. Nowadays, most reputable steam boiler manufacturers consider the piping immediately around the boiler to be a part of the boiler. They publish installation booklets showing the contractors how they *must* install these boilers. If a contractor doesn't follow the manufacturer's instructions, the boiler won't operate efficiently, and there's a good chance the manufacturer won't honor their warranty should you have a problem. So, insist on seeing the installation-and-operating manual beforehand. Have the contractor show you the correct near-boiler piping for the unit. Have him explain it all. And have him include in his contract a clause saying that he will install the boiler according to those instructions. This alone will scare away heating contractors who don't know what they're doing when it comes to old steam systems. A knowledgeable contractor will never object to this very reasonable request.

✓ Don't accept copper tubing for the boiler's supply piping. Copper expands and contracts much more than steel. Because steam piping can twist into some odd angles, the expansion of the copper often puts a lot of torque on the soldered joints. That twisting frequently causes the joints to come undone after a few years. And then you're on your own. Copper can also leach out into the boiler, shortening the boiler's life through dielectric corrosion. Proper steam piping calls for threaded steel pipe and fittings. The threads allow the steel pipe to twist without coming apart. Copper tubing is what the low bidder will offer. If you're planning to stay in your home for more than a year, insist on properly installed, threaded steel pipe.

✓ Modern steam boilers sometimes don't get along well with old chimneys. Make sure you ask every heating contractor who steps into your home about your chimney. This is for your safety and protection. An old chimney may need to be lined with stainless steel to keep the flue gases from condensing on their journey to the top. Condensing gases form an acid that can eat through the mortar and cause parts of your chimney to fall apart. This, of course, can lead to poor venting of the products of combustion, and potentially dangerous levels of carbon monoxide in your home. Carbon monoxide is deadly. If the contractor won't talk about your chimney - if all he cares about is the boiler sale - he is not a professional. Go no further with this person. And know that a stainless steel chimney liner isn't cheap. You'll risk your life if it's needed and you try to do without it, though.

✓ Steam pipes must be insulated to keep the steam from condensing before it reaches the radiators. In the old days, we used asbestos. Nowadays, we spend a lot of money removing the asbestos, and we rarely replace it. Most folks figure the heat isn't "lost" because it's still inside the house. But if the steam is condensing in your basement pipes, it won't be condensing in your bedroom radiators on the second floor. You'll be burning lots of fuel and you'll be miserably cold in some of those rooms. You know all of this by now. If your steam supply pipes are uninsulated, have the contractor include a price for new insulation. Or insulate them yourself if you'd prefer.

✓ Reputable boiler manufacturers also include thorough cleaning instructions in their installation-and-operating manuals. The contractor must follow these instructions if your new boiler is to make dry steam. Have the contractor show you these instructions, and have him write on the contract that he will follow them to the letter. Then make sure he does. It takes nearly a full day to properly clean an old steam system. There is no chemical or magic potion I know of that can make decades worth of dirt vanish. Don't accept shortcuts. And so that you're better able to compare prices, ask each contractor to give you the price of the cleaning as an addendum to the installation contract. If he tells you the boiler doesn't need cleaning, and that he can solve any problems with a few cans of chemicals after the job is done, he's a knucklehead.

✓ Have the contractor install a drain valve in the boiler's mud leg. The mud leg is the drum at the very bottom of the boiler, the place where sediment will gather. If you don't insist on this drain valve, you probably won't get it. Most contractors will leave the drain valve out to save a few bucks, but without it, you won't have a way to flush sediment from your boiler as the years go by. Dirty boilers have shorter life spans than clean boilers. How often do you want to buy this thing?

✓ If you're not going to do it yourself, get the contractor to inspect your air vents and replace them if necessary. You'll find these air vents on your radiators (if you have a one-pipe steam system) and near the ends of your mains (on both one- and two-pipe steam systems). Good air vents make a dramatic difference in system performance. Even a brand-new boiler will gobble fuel if the air vents are old and clogged. Be prepared to pay extra for the vents, and know that there's a definite range of quality when it comes to these things. Like all things in life, you get what you pay for. Buy well from a reputable supplier, and take comfort in knowing that this is one of the best investments you can make in your old steam system.

✓ Heating contractors buy their equipment from plumbing and heating wholesalers who handle specific brands of boilers. Since the wholesaler extends credit to the contractor, the contractor will usually buy what the wholesaler stocks. Because of this, you may find it difficult to get three or four competitive prices if you base the comparison on a particular manufacturer's boiler. That's because the quoting contractors buy from different wholesalers. If you ask for Brand A and the contractor's wholesaler stocks Brands B or C or D, he will try to sway you to the brand that he's used to installing. There's nothing wrong with this. Rather than try to level the playing field by having them all quote on a certain brand of boiler, or a certain model number, do this: Let each contractor quote on what he thinks will best heat your home, but have each contractor guarantee the results in writing. Have them say, for instance, that when they're finished, your house will heat evenly, and with no noise or squirting air vents. Some contractors will run for the hills when they realize you're buying results instead of just a boiler, but you're better off without those guys anyway. A good contractor will talk to you seriously about your system's problems and their solution. He'll most likely suggest things that go beyond a simple boiler replacement. There are people out there who have no problem guaranteeing results.

✓ And know that competent steam contractors generally charge more for a boiler replacement than others, but these guys deliver those wonderful results - increased comfort and fuel savings. They will *never* be the low-bidder, but they will do what they say they'll do, and I think they're well worth their price.

A very good place to meet many competent contractors is on the Wall at our website, www.HeatingHelp.com.

You can also ask questions at your local plumbing & heating supply house. Stop by at a time when they're not that busy (not first thing in the morning or late in the day) and ask if you can speak to the "Heating Guy" (most places have one). He won't be selling you anything because his customer is the contractor, but he'll be able to tell you about the good guys in the area.

Finally, know that a good steam contractor doesn't have to be old. Some of the sharpest steam contractors I know aren't even 40 years old. Each of them has done the research, though, and each has a passion for this stuff. *That's* the guy you're looking for.

Good luck, and remember, you are *never* alone, no matter where you live. We're always as close as the World Wide Web. Stop by and visit!

www.HeatingHelp.com

Dan Holohan

Hug your kids.

INDEX

NOTES